Tim Lindstrom, a committed conservationist, decided to run for the Virginia state legislature in 1991.

League of Conservation Voters: "What was the difference between you and your opponent?"

Lindstrom: "I was perceived as a moderate who was committed to conservation.... My opponent was perceived as a very, very conservative but decent guy.... He came out against any new regulations on land."

LCV: "What were your chances?"

Lindstrom: "The district I was running in was extremely conservative, and my opponent outspent me by about two to one. So I really didn't think I was going to win. In fact, with about four weeks to go in the campaign, a poll showed I was losing by about 2 to 1."

LCV: "What happened on election day?"

Lindstrom: "According to reports after the polls closed, the election had ended in a tie. But the next day when the official results were released, it turned out I had won...by *one vote*. Then, after a recount, it turned out I had lost by seven votes."

LCV: "So only eight votes would've made the difference?"

Lindstrom: "That's right."

ABOUT THE LEAGUE

The League of Conservation Voters (LCV) is the 21-year-old non-partisan political arm of the national environmental community. Our goal is to change the balance of power in the U.S. Congress to reflect the pro-environment concerns of the American public.

WHAT WE DO

• Each year, we publish the *National Environmental Scorecard*, which details the voting records of members of Congress on important environmental legislation. We distribute it to our members, the media, and concerned voters nationwide so that representatives can be held accountable for their actions on election day.

• LCV also publishes *Presidential Profiles*, which provide in-depth analyses of the candidates running for president.

• Periodically, we release Congressional committee scorecards, which highlight committee members' votes on significant pieces of legislation.

• The League also works to elect environmentalists to the U.S. Senate and House of Representatives—because legislation that truly protects our nation's natural resources will *only* be passed when there is a pro-Earth majority in Congress. We provide political support to campaigns, including placement of trained campaign staff, media assistance, and direct financial contributions.

JOIN US!

For information about joining the League, write to:

League of Conservation Voters - Membership
1707 L Street, NW
Suite 550J
Washington, DC 20036

VOTE
FOR THE
EARTH

**THE EARTHWORKS
GROUP
&
THE LEAGUE OF
CONSERVATION
VOTERS**

EARTHWORKS PRESS
Berkeley, California

THIS BOOK IS PRINTED ON RECYCLED PAPER

Created and Packaged by Javnarama
Designed by Javnarama

ISBN 1-879682-08-7
First Edition, 7/92
10 9 8 7 6 5 4 3 2 1

We've provided a great deal of information in this book. We've relied
on advice, recommendations, and research from others whose
judgments we consider accurate and free from bias. However, we
can't and don't guarantee the results. This book offers you a start.
The responsibility for using it ultimately rests with you.

For information on bulk rates
and customized editions, write to:

EarthWorks Press
1400 Shattuck Avenue, #25
Berkeley, CA 94709
(510) 652-8533

ACKNOWLEDGMENTS

The League of Conservation Voters and The EarthWorks Group would like to thank everyone who helped make this book possible. (Organizations listed below for identification purposes only.)

- Bruce Babbitt
- Jim Maddy
- Ali Webb
- Claudia Schechter
- John Javna
- Sven Newman
- Melissa Schwarz
- Catherine Dee
- K Kaufman
- Fritz Springmeyer
- Lyn Speakman
- Nancy Skinner
- Lenna Lebovich
- Jack Mingo
- Kim Koch
- John Dollison
- Joanne Miller
- Emma Lauriston
- Dayna Macy
- Denise Silver
- Andy Sohn
- Sharilyn Hovind
- Dan Irelan
- S. M. Redel
- Roger Stephenson
- Elaine Lynch Jones
- Nina Tracy
- Hedrick Belin
- Donna Koren
- Sarah McCourt
- Bronwyn Reynolds
- Laura Deame
- Joyce Letendre
- Nancy Rollman
- Jay Daly
- John Demos
- Edith Rose
- Ronda Copher
- Timothy Deal
- Scott Moore
- Ross Newcombe
- John Savage
- Suzanne Thomas
- Stuart Wolphaupter
- H. Kevin Eddins
- Kelly Jones
- Rebekah Lacey
- Laura Porio
- Bob Price
- Nathan Tyler
- Michael Silverstein
- Denis Hayes
- Bob Rose
- Tim Lindstrom
- Sally Ethelston
- Steve Moyer
- Joe Goffman
- Tim Searchinger
- Lucy Blake
- Ray Barry
- Jim Aidala
- Seth Zuckerman
- Bill McKibben
- Michael Bean
- Campbell Plowden, Greenpeace

- Peter Montague, Environmental Research Foundation
- Jeffrey Tryens, Center for Policy Alternatives
- Liz Cook, Friends of the Earth
- Allen Rosenfeld, Public Voice for Food and Health Policy
- Jon Kusler, Association of State Wetland Managers
- Pesticide Action Network
- Chris Flavin, Worldwatch Institute
- Alden Meyer, Union of Concerned Scientists
- Nancy Hirsh, Energy Conservation Coalition
- Pam Welher, Rainforest Action Network

American Oceans
- Andy Palmer
- Gari Bodor

Center for Marine Conservation
- Marydele Donnelly
- Gary Magnuson

Environmental Action
- Lisa Collaton
- Nancy Hirsch

Greenseal
- Norman L. Dean
- James B. Dougherty

National Audubon Society
- Clark Williams
- Lisa Glantz
- Brock Evans

National Coalition for Alternatives to Pesticides
- Jay Feldman
- Jane Kochersberger

Natural Resources Defense Council
- Chris Calwell
- David Doniger
- Jane Mardock
- Debbie Sheiman
- Faith Cambell
- Jessica Landman
- David Driesen
- Erik Olson
- Bob Adler
- Dan Reicher

Sierra Club
- Bill Mankin
- Daniel Weiss
- Nancy Wallace
- Cathy Fogel
- Dan Becker
- Steven Krefting

U.S. Public Interest Research Group
- Gene Karpinski
- Carolyn Hartmann
- Rob Stuart
- Dave Hamilton

Wilderness Society
- Steve Whitney
- Nancy Green

World Resources Institute
- Rafe Pomerance
- Sean Fox

Zero Population Growth
- Susan Weber
- Nancy Hirshbein
- Dianne Sherman

CONTENTS

Environmentalism: A Definition..9

THE ENVIRONMENT AND POLITICS
Every Candidate Counts..14
The Environmental Connection..17
The Economy & the Environment...21
Who Are The "Pro-Earth" Candidates?".......................................24
Buzzwords...26

15 REPRESENTATIVE ISSUES
Protecting the Planet
Vote to Stop Global Warming..29
Vote to Stop Ozone Depletion..33
Vote for the Rainforests..37
Vote to Save Our Oceans..41
Vote for Family Planning..45

Saving America
Vote to Control Toxic Waste...50
Vote for Wetlands...53
Vote to Save Endangered Species..57
Vote for Energy Efficiency..62
Vote to Save Ancient Forests..67

Helping at Home
Vote for Recycling..73
Vote for Safe Food...77
Vote for Clean Air...81
Vote for Clean Drinking Water..85
Vote for "Green" Products...89

CANDIDATE RECORDS
The 1991 National Environmental Scorecard..............................94
Find Out What's Going On in Your State....................................120

ELECTION RESOURCES..132

My friends care deeply about the Earth, but they have given up voting for it. Instead, they put their energy into reducing and recycling their trash, insulating their houses, and shopping for products certified by Green Seal.

"Why," they ask me, "should I waste my time working on elections?"

The answer is simple: If we don't start electing environmentalists to office, we will lose the world. Like it or not, government sets and enforces the rules that determine how we live and work. To ignore government is an indulgence we can't afford.

Recycling, by itself, can't prevent global climate change. Street demonstrations, by themselves, can't clean up toxic waste dumps. Being a green consumer, by itself, can't save the Florida Everglades or the Alaskan National Wildlife Refuge. *There are many important things that only government can do.*

The real question is not, "should we vote?" but rather, "who should we vote for?"

Politicians can fool all of the people some of the time. How do you avoid being taken for a fool?

Read this book—learn about the issues…find out what our elected officials are doing…find out what they could be doing. And then VOTE!

Denis Hayes
Chairman of Earth Day

ENVIRONMENTALISM:
A DEFINITION

Are you an environmentalist? What does that mean to you? To some people it's recycling regularly. To others, it means joining a carpool or becoming a member of an environmental group. Some people consider themselves environmentalists when they pick up litter.

When "environmentalism" is personal, we don't need to have a definition for the term. But in politics, when one individual can be speaking to—and for—millions of voters, a definition is critical.

Right now, we don't have one. When politicians declare that they're "for the environment," you can't really be sure what they're saying.

TAKING ADVANTAGE

This confusion makes it easier for anti-environmental politicians to pass themselves off as "pro-Earth." Since there's no standard—no "accepted" definition—for voters to hold politicians to, they can get away with making any claims they choose. Dan Quayle, for example, who's done as much as any other individual to *undermine* environmental regulations in the Bush Administration says he's committed to protecting the environment because he takes his kids hiking and fishing.

Ironically, once voters accept this hollow definition of environmentalism, candidates who are truly committed to protecting the environment start to look bad by comparison. They're perceived as "Alarmists," ..."Extremists"... "Anti-human"... "Impractical." Rhetoric wins out over reality.

DEFINING TERMS

To stop this travesty—and establish environmentalism as a real force in the political arena—we need a clear, simple definition of what the term *environmentalism* means.

This definition has to be broad enough to be meaningful to all

Americans who sincerely consider themselves environmentalists. And it must show that environmentalism is undisputably pro-human, practical, and mainstream.

Fortunately, that's easy to do.

LIFE SUPPORT

In its broadest sense, environmentalism deals with survival. If we poison our air, our water, our food...if we destabilize the climate and destroy the atmosphere...we—and most other living things—will die.

You can't get a much more practical, pro-human concern than that.

So to put it into one sentence: In politics, environmentalism is protecting the life support system of the planet.

NAYSAYSERS

Inevitably, someone will say, "You're being alarmist. Our life support system isn't at risk. There's no threat." But the international scientific community disagrees. They've gathered an overwhelming amount of data that shows clear threats to:

√ Our air. Because of pollution, more than a billion people worldwide already live in places where *breathing* is actually hazardous to their health.

√ Our drinking water. The EPA estimates that 50% of America's drinking water already has some level of contamination.

√ Our food. Overfishing is reducing the fish population drastically; poor farming and logging practices are helping erode between 2.7 and 3.1 billion tons of topsoil from U.S. cropland every year; a 2-year FDA study found that 70% of U.S. seafood was contaminated with pesticides; etc.

√ The atmosphere. Global warming is destabilizing our climate, and the disappearance of the ozone layer threatens to increase cancer and destroy the food chain.

Clearly, environmentalists' concern for the Earth's life support system is not only justified, but an urgent priority. That's why we've got to stay focused on the meaning of the word...and elect more candidates who believe in it.

WHO IS AN ENVIRONMENTALIST?

Now that we've got a political definition for the term environmentalism, we need a political label for the people who believe in it.

Why a political label?

For one thing, much of American politics seems to depend on quick, easy identification of a candidate's stand on the issues.

For another, it's important to establish that protecting the life support system of the planet is nonpartisan and non-ideological. Republicans & Democrats...traditional liberals & conservatives...all concerned Americans need to feel welcome in the environmental camp.

But perhaps the most important reason is that when we don't label ourselves, our opponents do instead. They've called environmentalists tree-huggers, left-wing radicals, etc. You've heard the terms. Their main purpose is to direct voters *away* from the real issues. We need a term that directs voters *back*.

THE ECO-CONSERVATIVE

Although some people might initially be uncomfortable with the idea, the term "conservative" rightfully belongs to environmentalists...because the root of the word is "conserve"—i.e., to resist spending; to avoid waste. In fact, the dictionary definition of the term conservative is "someone who is cautious about change."

A person whose primary political mission is to protect the life support system of the Earth will always be cautious about making changes that might damage it. He or she will weigh every issue in terms of its impact—or potential impact—on human survival first...and other considerations second. And if there's any doubt about the outcome, the environmentalist will err on the side of protecting our life support system—not undermining it.

That person is an "Eco-conservative."

WHAT IT MEANS

An eco-conservative is, to the resources of the planet, what a fiscal conservative is to money.

• Just as fiscal conservatives are cautious about using capital, eco-conservatives are cautious about depleting resources that support life on the planet.

- Just as fiscal conservatives believe their money belongs in strong, long-term investments, eco-conservatives believe in action that strengthens the planet's ability to support life in the long-term.
- Just as fiscal conservatives consider it a bad risk to "borrow against the future," eco-conservatives are unwilling to improve our own quality of life at the expense of future generations.

WHO'S THE RADICAL?

If the concept of "eco-conservatism" seems surprising, it's probably because in the last few decades, radical anti-environmentalists have co-opted the term, and managed to convince the American people that they're actually "conservatives." But consider this—which is more *conservative* and which is more *radical*?

√ Managing a forest so that it will produce timber indefinitely, support wildlife, prevent flooding and drought, and provide recreation…or cutting down every tree—leaving behind barren land that can't sustain wildlife, retain topsoil, or provide jobs?

√ Use less energy to fight global warming, protect the oceans from oil spills and offshore drilling, reduce toxic waste and air pollution, and preserve valuable resources for future generations…or keep using more and more energy—sapping our limited oil reserves, despoiling wilderness, polluting the air and the ocean, creating more toxic waste for our children to deal with?

√ Maintaining wetlands so they continue to combat water pollution, protect us from floods, and provide spawning grounds for fish that our seafood industry relies on…or fill them in—and then frantically spend billions of dollars to replace the water purification and flood control they provided, while the seafood industry gradually declines?

OUR CHOICE

As you'll see in this book, the notion that Americans have to choose between a high standard of living and environmental protection is a dangerous fantasy.

The simple truth is that in today's world, eco-conservatism is the *foundation* of a high standard of living…and a healthy economy…and a strong America…and every other important social and economic issue. It's up to each of us to take that truth into the voting booth this year…and elect people who understand it.

THE
ENVIRONMENT

AND
POLITICS

EVERY CANDIDATE COUNTS

At every level—national, state, and local—the people you vote for can help…or harm…the environment. Here are just a few of the thousands of examples that prove the candidates we select do make a difference.

THE PRESIDENCY

A president with vision can have a positive effect on federal environmental policies for generations.

• President Teddy Roosevelt, a conservationist, backed a law in 1906 that gave future presidents the power to protect natural resources. Seventy-five years later President Carter used the law to protect much of Alaska by making it a wilderness preserve.

A president with limited commitment to environmental protection can have a powerful impact, too.

• In 1991, President Bush proposed a new federal definition for the term "wetlands," which increases the number of days that land has to be wet to qualify for protection. If the new definition is adopted, it could result in the loss of almost half of our remaining wetlands.

THE CONGRESS

Individual senators and representatives can make a difference by creating legislation to protect a sensitive part of the environment…

• Senator Dave Durenberger of Minnesota made cleaning up contaminated drinking water one of his top priorities. In 1984, he helped pass legislation to prevent, detect, and clean up leaks in underground gas tanks. Two years later, he passed a law that requires tank owners to have insurance so the cost of cleaning up leaks will always be covered.

• Indiana Rep. Jim Jontz has made a personal commitment to saving our ancient forests. So far, his Ancient Forest Protection bill has 134 co-sponsors.

• In 1990, to encourage food companies to sell "dolphin-safe" tuna, Rep. Barbara Boxer of California helped pass a law requiring food companies to inform consumers, on the label of each can, what kind of net had been used to catch the tuna. Rather than admit to consumers that it was killing dolphins with its fishing nets, the tuna industry changed the way it fishes.

...or to damage the environment.

• In 1991, Senator Bennett Johnston of Louisiana introduced a bill to open up 1.5 million acres of the Alaskan Wildlife Refuge to oil and gas drilling. Fortunately, the bill was blocked by a group of senators who were elected, in part, because of their pro-environment stance (and with the help of the League of Conservation Voters.)

STATE OFFICES

State officials help pass environmental laws that are much stricter than federal laws, paving the way for tougher federal laws in the future.

• In 1989, Massachusetts Rep. Geoffrey Beckwith pushed through the first "pollution prevention" law in the U.S. designed to help companies reduce their use of toxic chemicals. Other states are now following his lead.

• By executive order, Governor Thomas Kean of New Jersey created the nation's first energy-efficiency program specifically designed to fight the greenhouse effect (global warming). His program requires all New Jersey State agencies to reduce emissions of carbon dioxide, CFCs, and other gases, and promotes tree planting.

• Under a law written by Kurt McCormack, a Vermont state legislator, cars with air conditioners that use CFCs will be banned in that state starting in 1993. This is currently one of the toughest laws on CFCs.

And if the federal government won't address the special needs of your region, it's often up to state officials to step in.

• Hazardous waste is an especially serious problem in Georgia, but the federal government wasn't taking care of it. So Governor Zell Miller created a state-level "Superfund" law to clean up 800 toxic dump sites. Ninety-eight percent of those sites would have received no cleanup funds from the federal government.

LOCAL OFFICES

Local elected officials are constantly making decisions on environmental issues that affect the quality of your immediate environment.

• The mayor and town council of Woodland, North Carolina, passed special zoning laws to allow a hazardous waste company to build a new incinerator in Woodland. In the next election, angry citizens elected only anti-incinerator candidates, including Mayor Bill Jones, who restructured zoning laws to prevent incinerators from being built in town.

• In 1990, Mayor Jim Donchess of Nashua, New Hampshire, led a citizens' revolt against a local developer who planned to build on wetlands in the area. The mayor called a special election, and a proposition was passed that created a "no-net-loss" policy for local wetlands.

...But local officials also make decisions that affect the entire country.

• When the city of Santa Barbara faced a severe water shortage, Mayor Sheila Lodge had a choice: she could develop a citywide water conservation program, or spend city funds to transport water from distant sources. She decided to conserve, and the city reduced its water use by 45%. Her program now serves as a model for other drought-prone cities.

• Mayor Sharpe James of Newark, New Jersey, decided to make the Newark recycling program one of the most ambitious in the country. It includes curbside recycling in both residential and commercial neighborhoods and requires city agencies to buy recycled products. The nationally recognized program is responsible for a 50% reduction in Newark's garbage.

• In 1989, Mayor Larry Agran of Irvine, California, decided to do something to protect our ozone—He imposed a city ordinance restricting the use of ozone-depleting clorofluorocarbons (CFCs). Other cities around the country soon did the same. These laws led to national action: in 1992, President Bush ordered an accelerated phase-out of CFCs.

THE ENVIRONMENTAL CONNECTION

Most of us believe environmental issues are separate from more traditional political concerns, like jobs, national security, and family values.

But the environment is actually an integral part of *all* major political issues. That means if you're not taking the environment into consideration when you vote, you're not fully addressing *any* of the issues. Here are some examples.

HEALTH CARE

Our Concern: America's health-care system is deteriorating, while costs skyrocket. Today, $1 of every $9 spent in the U.S. goes to health care. Yet health care is still not adequate: one of every seven Americans has no medical coverage—including eleven million children.

The Environmental Connection: The pressures making our health-care system less efficient and more expensive are caused, in part, by environmental problems. Polluting the Earth poses severe health risks. For example:

• Cancer rates are soaring—and, according to the EPA, about 1 in 5 pesticides used on food in the U.S. may cause the disease.

• The EPA predicts that, if ozone depletion isn't stopped, over the next 50 years as many as 12 million more Americans will get skin cancer.

• According to the American Lung Association, air pollution costs Americans as much as $100 billion a year in health costs and lost productivity.

• About 25% of prescription medicine is derived from rainforest plants—yet every minute of the day, 80 acres of rainforest are destroyed, wiping out the potential for new "miracle" drugs.

NATIONAL SECURITY

Our Concern: Protecting the country from foreign threats is a primary goal of U.S. foreign policy.

The Environmental Connection: Our failure to conserve energy has made us vulnerable to the pressures of foreign governments. Because nearly 50% of our oil is imported, we must rely on other countries for our energy survival.

Yet, instead of reducing this dependence by becoming more energy efficient, we're planning to use more energy—and more oil. President Bush's new "National Energy Strategy" calls for the United States to consume 37% more energy in 2010 than it's using today—including 12% more oil. And the president's 1993 budget reduces energy conservation programs by $15 million.

JOBS

Our Concern: Unemployment is at 7.8%, an 8-year high. Some 16 million people are either unemployed or underemployed. Where are new jobs going to come from?

The Environmental Connection: According to Christopher Flavin, of Worldwatch Institute, "green" industries can provide them. For example:

• "Pollution controls spur economic activity and create jobs. An analysis of 1988 U.S. pollution control expenditures suggests that the industry supported nearly 3 million direct and indirect jobs. Some 2.5% of U.S. jobs are now in pollution control, almost all of them created since 1970. Studies project these jobs will continue to expand rapidly in years ahead."

• "Recycling will create more jobs, since it is much more jobs-intensive than the alternatives—landfilling or [incinerator] projects. A study in Vermont found that for each one million tons of materials processed, recycling generates between 550 and 2,000 jobs."

• "Energy development is another field ripe for job growth. An industry study shows that investing in energy conservation and renewable energy sources has the potential of generating 175,000 new jobs by the year 2010."

CIVIL RIGHTS

Our Concern: Racial minorities are still not being treated fairly.

The Environmental Connection: Minorities suffer a disproportionate exposure to environmental hazards. According to one recent report, for example, "Communities with commercial hazardous waste facilities had twice the percentage of racial minorities as communities without such facilities. Communities with more than one facility, or with the largest toxic waste dumps, have three times as many people of color."

TAXES

Our Concern: Americans are tired of seeing tax dollars wasted on unnecessary expenses and subsidies of big business, while essential services like police, firefighting, and education deteriorate.

The Environmental Connection: Environmentally irresponsible business practices cost taxpayers billions of dollars every year. We shouldn't have to pay for their cleanup. For example:

• Cleaning up all of America's hazardous waste sites will cost taxpayers at least $1 trillion.

• Every year, water pollution cleanup costs Americans an estimated $10 billion.

• Disposing of the country's trash—including many things that could be recycled—costs taxpayers at least $4 to $5 billion a year.

• In addition, taxpayers subsidize businesses like ranching, mining, and logging when our public lands are leased to private citizens at below market cost. For example: grazing cattle on U.S. lands costs cattle ranchers only a tenth of what it costs them on private lands. Taxpayers make up the difference.

FAMILY VALUES

Our Concern: As a society, we need to reaffirm our commitment to the family.

The Environmental Connection: The phrase "family values" is being used a lot in this election year—but not in conjunction with the environment. That's a mistake, because preserving the Earth for our children is at the heart of the issue. How could anyone be *for* families, for example, and not be for clean water, clean air, and safe food? Here are a few of the many ways that family values and the environment are linked:

• *Safeguarding our children's health:* Air pollution, water contamination, pesticides, toxic waste, and ozone depletion all threaten children, who are most vulnerable to their effects. The worse environmental degradation gets, the more likely it is to harm them.

• *Feeding our families:* Traditionally, this means earning money to "put food on the table." But it's getting even more basic. Poor farming and logging practices are eroding our topsoil; overfishing is reducing the fish population; global warming, ozone depletion, and air pollution are affecting world crops. Americans may even experience food shortages some day soon if we don't take action now.

• *Teaching our children right and wrong:* Many people consider instilling a sense of right and wrong in children the cornerstone of "family values." Clearly, treating our planet with care and respect is the right thing to do; it should be taught by example. In this context, environmentalism is a moral imperative—as well as a practical—choice.

• *Protecting our children's future:* We'd all like to leave our children—and their children—a better life than we had. This means two things: a world they can enjoy, and a world that will sustain them. In terms of the environment, "enjoyment" means clean rivers and streams, healthy forests, and unpolluted beaches. But as a survival issue, it means breathable air, a stable climate, an undamaged ozone layer, and so on. It's hard to believe we've reached the point where our great-grandchildren might not have the most basic elements of life. But that's where we are. If you care about family values, vote to protect the Earth.

THE ECONOMY &
THE ENVIRONMENT

This piece was contributed by Michael Silverstein, president of Environmental Economics, a Philadelphia consulting and research firm.

For decades, the economy has been the Achilles' heel of the environmental movement. This is because most politicians, polluters, and even some environmentalists have accepted the idea that we must choose between economic growth and environmental protection.

This idea that a choice is necessary is so ingrained in the public consciousness that people cling to it—in spite of the fact that it's now totally outdated.

The truth is, our economy can only stand to *gain* from environmental protection. The facts speak for themselves.

JOBS

Environmental legislation and regulation does *not* reduce the number of jobs available to American workers. It increases them. For example:

• At present, an estimated two million Americans earn a living doing some kind of environmental cleanup work. Another million are indirectly employed by this industry.

• Some 65,000 to 70,000 companies are active in this field.

• The 40 largest U.S. environmental companies alone employed more than 146,000 people at the end of 1991. Their average payroll increased 8.6 percent last year, at a time when thousands of companies in industries like retailing, banking, auto manufacturing, and computers were laying off staff.

• It is true that some jobs will be lost as the U.S. and the world economy become more environmentally sensitive. But, just as when this country went from the horse and buggy to the car, new jobs will also be created. Some people *will* have to change careers, but trying to hold onto jobs that are becoming obsolete is only delaying the inevitable.

• For example, the logging industry claims that protecting the ancient forests in the Northwest would cost thousands of jobs. But, with current logging practices, those forests will be destroyed in 10 to 20 years, leaving loggers with no forests *and* no jobs. It makes sense from both an environmental and an economic standpoint to take pro-active steps now to assist loggers in switching to fields that are growing.

INNOVATION & TECHNOLOGY

Environmental protection does *not* inhibit American business innovation and technology. Instead it encourages it. For example:

• An estimated 1,500 technologies (recycling, emissions control, energy efficiency, etc.) are involved in the fight to save the world ecosystem.

• Innovations in the environmental protection field are quickly taking the place of defense and aerospace in spurring technical advances, which in turn create new jobs and profits throughout the economy.

INTERNATIONAL COMPETITION

Laws aimed at protecting the environment have *not* made the United States less internationally competitive. The opposite is true.

• Polluting by industries is a sign of inefficiency and poor management. The more efficient a factory is, the fewer raw materials are lost in emissions and waste. By allowing businesses to avoid pollution regulation, all the government is really doing is temporarily insulating inefficient producers from the need to innovate and invest in new equipment.

• The same laws and regulations that force U.S. businesses to reduce pollution also make them more efficient and better able to compete with foreign manufacturers.

• One example of an American company that has turned environmental protection into a profit is 3M: A decade ago 3M initiated the PPP program (Pollution Prevention Pays). The idea was to reduce pollution and change production in ways that would help the company and protect the environment. To date, the company has saved itself more than $600 million.

CLEANUP

Cleaning up the world's environment in coming years shouldn't be seen as a *problem*. It is an economic *opportunity* the U.S. can't afford to pass up. For example:

• Official projections estimate that $1.2 trillion will be spent in the U.S. during the 1990s for environmental cleanup.

• A total of $3.5 trillion will be spent worldwide. From a purely ecological perspective, it doesn't matter whether this money flows into the coffers of Germans and Japanese, Swiss and Swedes, Angolans and Laplanders.

• But it matters a great deal to the future of the U.S. economy. We lost our dominance in the world car market…in the computer field…in personal electronics. To lose our lead in the environmental protection market, at a time when international spending in it is soaring, would be a national economic disaster.

VOTER ALERT

The issue for American voters when it comes to a healthy environment *and* a healthy economy is not to choose one or the other. In today's world, you either have *both* or *neither*.

• Candidates who show an understanding of the link between environmental protection and job production should be rewarded at the polls.

• Those who still prefer to defend the short-term interests of polluters, and who still think they are "protecting American jobs and profits" when they fudge on the environment, should be put out to pasture.

WHO ARE THE "PRO-EARTH" CANDIDATES?

I n 1988, candidate George Bush promised to be the "environmental president." But once he was in office, President Bush cut energy conservation programs, asked Congress to approve oil drilling in the Arctic Reserve, helped companies circumvent his own Clean Air Act, and pushed for legislation that could destroy as much as half of our remaining wetlands.

This election year he's claiming to be an environmentalist again.

That's not surprising. Politicians go where the votes are, and a recent poll showed that the environment will be a "major issue of concern" for some 60% of American voters in November.

Since so many candidates will be running as "environmentalists," it's important for voters to be able to tell which ones really mean it. Here are some simple ways to check them out.

1. Check Their Records
• You can learn the most about an incumbent's position by examining their voting record on environmental issues. To help you do this, the League of Conservation Voters publishes the *National Environmental Scorecard* every year to help voters evaluate Congress's environmental decisions. Part of it is included in the back of this book. You can order the complete "Scorecard" from LCV at 1707 L Street, NW, Suite 550J, Washington, DC 20036. The cost is six dollars (checks only). You can also call (202) 785-8683 for information on specific candidates.

2. Ask the Experts
• Evaluating non-incumbent candidates is a little more difficult, since they don't have voting records to examine. However, environmental groups can generally help you evaluate local and national candidates by providing background information on them. For example: what is a candidate's employment history? Has he

or she worked for companies with poor polluting records? Also, find out which candidates have been endorsed by environmental groups.

• Eco-political organizations from every state are listed in the back of this book. Most can provide you with information about individual candidates in your state.

• The Sierra Club's local chapters around the country offer specific information about candidates at all levels (local, state, and national) and will tell you which candidates they endorse. To find the chapter nearest you, call the Sierra Club's public information number: (415) 776-2211.

3. Ask Questions

• Do your own candidate evaluations as you listen to speeches and debates, read newspapers and campaign literature, etc.
For example:

√ How long has the candidate been working for the environment? Is it a real commitment, or have they just jumped on the eco-bandwagon?

√ Does the candidate work with environmental groups?

√ What bills or policies related to the environment has the candidate introduced, sponsored, or supported?

√ Does the candidate talk about environmental issues consistently, or only when addressing environmental voters?

Note: When in doubt, ask what a candidate is willing to *spend* to achieve environmental goals.

4. Make a Phone Call

• If you want to know where a candidate stands on specific environmental issues, call or write their campaign offices and ask for copies of their position papers. If a candidate has no position paper on an issue, that may tell you something.

5. Beware of Buzzwords

• Watch out for catchphrases coined to make candidates sound pro-environment when they really aren't. For a list of some buzzwords being used in this election, check out the following page.

BUZZWORDS

Some of these phrases are meant to elicit an emotional response from listeners—even though they don't really mean anything. Others are downright deceptive. Here are a few you're sure to hear candidates say some time during this year's political campaign.

Acceptable Risk. Translation: Willing to sacrifice *other* people's health to avoid regulating companies.

Balance (or Balanced Use). Sounds good, but what it really means is: "Let's not go too far with this environmental stuff."

Flexible (or Cost-Effective) Solutions. Code words for weakening environmental regulations.

Jobs vs. Environment. A favorite with anti-environmentalists because so many people believe it—even though it's a lie.

Level Playing Field. Usually used by an industry when begging for more government subsidies, tax breaks, and exemptions from environmental regulations.

Multiple Use. Translation: Multiple abuse. Usually means sacrificing an environmentally sensitive area.

Protect the Free Market. Used when a candidate is protecting massive government subsidies for a polluting industry or business.

Responsible. Sounds positive, but it's often used by corporate fronts ("Citizens for Responsible ____").

Restrictive (or Costly). An emotional way of attacking environmental regulations.

Streamline Environmental Regulations. Sounds efficient, but it often means regulations will be weakened.

Sustainable. People on both sides of the issues use it, so context is important.

Wise Use. A term adopted by people who support logging, mining, off-road vehicle use, and grazing on public lands and wilderness. More accurately refered to as "waste and abuse."

15 REPRESENTATIVE ISSUES

To demonstrate how the environment and politics are linked, we've selected and explained 15 environmental issues that will be affected by the coming election. All are important, but they are only a sampling—picked to show you some of the ways the politicians you vote for can have an impact on the planet.

We've also given you an idea of what people against environmental protection are saying about these issues, and have provided a few answers to their arguments.

Finally, we've included tips on how to tell where candidates stand on these issues, to help you make your decisions at the polls.

The rest is up to you.

PROTECTING

THE PLANET

VOTE TO STOP GLOBAL WARMING

*All but one of the 10 hottest years on record worldwide
have occurred in the last decade.*

The greenhouse effect—when it's functioning normally—keeps our planet warm. Certain natural gases, like carbon dioxide, form a transparent "blanket" in the atmosphere that lets sunlight reach the Earth's surface, but keeps heat from escaping (like the glass in a greenhouse). This gas "blanket" traps heat close to the surface and warms the atmosphere.

But now, because of emissions from cars, power plants, and other sources, the "greenhouse blanket" is getting thicker. Average global temperatures are already nearly one degree higher than they were 100 years ago, and scientists predict increases of anywhere from three to eight degrees in the next century. This may not sound like much, but it only took a rise of a few degrees to end the Ice Age. Global warming could result in flooding, drought, and other environmental disasters.

The United States produces more greenhouse gases than any other country in the world. If we want to stop global warming, everyone in America—individuals, businesses, and government—will have to work together.

DID YOU KNOW

• "Greenhouse gases" include carbon dioxide (CO_2), produced by burning fossil fuels (coal, natural gas, and oil) and cutting down trees; chlorofluorocarbons, used in air conditioners and refrigerators (see Vote to Stop Ozone Depletion, p. 33); methane; nitrous oxide; and ozone.

• "Greenhouse gases" don't just disappear—gases emitted today will contribute to global warming for the next 100 years.

• CO_2 is the main greenhouse gas. It's responsible for 60% of global warming.

• As a result of deforestation and burning fossil fuels, the atmosphere now contains 25% more CO_2 than it did just a century ago.

• Although the U.S. has only 5% of the world's population, it produces 23% of global CO_2.

• American cars and trucks are responsible for about 30% of our CO_2 production. For every gallon of gas a car burns, it puts about 20 pounds of CO_2 into the air.

• If the U.S. continues its current energy-use patterns, our CO_2 emissions will nearly *double* by the year 2030.

WHAT OUR POLITICIANS ARE DOING: *A few examples*

The White House. Since his election in 1988, President Bush has consistently refused to commit the U.S. to reducing—or even stabilizing—CO_2 emissions, despite the fact that almost all other industrialized nations have made this pledge. In spring 1992, his threatened boycott of the Rio Earth Summit resulted in a watered-down global warming treaty, with no firm targets or dates for CO_2 reductions.

Congress. Attempts to pass a bill requiring the U.S. to stabilize CO_2 emissions at 1990 levels by the year 2000 were undermined by the Bush administration and a lack of congressional leadership. To get such a bill through, we need more environmentalists in the House and Senate.

The World Bank and **International Monetary Fund.** These bodies should encourage energy efficiency and other development policies that would help reduce CO_2 in Third World countries. Instead, they support and fund development projects that increase the use of fossil fuels. The White House sets the agenda and Congress approves the funds—so they both influence this policy.

State Governments. Some states have passed anti-global warming laws and goals. For example: Oregon plans to cut its CO_2 emissions by 20% from 1988 levels by the year 2005; Vermont plans a 15% reduction by 2000; and California is considering a similar goal.

Local Governments. The Urban CO_2 Project, an international consortium of cities (which includes Denver and Minneapolis in the U.S.), was formed in 1991. Its purpose: Cut greenhouse gas emissions by 20% in each city. Local governments can develop similar programs or use this project as a model.

THE POLITICAL DEBATE: *Some arguments you might hear*

Argument: "There's no proof that the greenhouse effect is real."

Answer: That's what cigarette companies said about the link between cigarettes and cancer…and we know how wrong they were. People who discount the greenhouse effect often have a stake in burning fossil fuels—like coal and oil companies.

But the fact is, the basic science underlying the theory of the greenhouse effect is indisputable—greenhouse gas concentrations *are* increasing in the atmosphere, and the Earth *will* get warmer. Now scientists are trying to answer two questions: How hot will it be, and how fast will it occur? The problem is, by the time we find the answers to these questions, we may have already committed the world to changes that are potentially catastrophic.

Argument: "We don't know how global warming will affect our world. We should do more research before taking any action."

Answer: For fifteen years, panels of the world's top scientists have consistently concluded that a doubling of greenhouse gas concentrations will result in a 3- to 8-degree increase in the Earth's temperature. If this occurs in the next century (as predicted), the change would be the most dramatic climatic shift since the Ice Age. Though scientists cannot yet predict exactly how certain regions will be affected by the change, scientists around the world warn that unless we act soon, the world will experience more serious drought, a rising sea level, and widespread ecosystem damage. And, though everyone agrees that scientific research must continue, there is no question that we know enough now to act. Putting off cost-effective steps to reduce greenhouse-gas emissions is like refusing to take out a free insurance policy on the Earth.

Argument: "Stabilizing greenhouse-gas emissions would hurt the economy."

Answer: When politicians put short-term economic interests over the long-term health of our economy and environment, we all lose.

The Bush administration has opposed a stabilization target for greenhouse gases, even though the government's own analysis demonstrates that the U.S. could reach such a target simply by pursuing energy efficiency and conservation measures. Most other western countries, such as Germany and Japan, have already agreed to stabilization.

Moreover, a study done by the four leading energy/environmental groups shows that if our country invested enough in energy conservation and efficiency measures, the U.S. could reduce its carbon dioxide emissions by 70% by the year 2030 and earn a net savings of $2.3 trillion at the same time.

VOTE FOR THE EARTH: *Picking the right candidates*

√ Do they believe that global warming is a problem?

√ Are they willing to actively support legislation to stabilize or reduce CO_2 emissions?

√ Would they support an energy or carbon tax aimed at helping reduce CO_2 emissions in the U.S.?

√ Do they agree that state and local governments have a role in solving this problem?

√ Do they have concrete proposals to cut CO_2 emissions locally as well as nationally?

Voter Notes

• Watch out for candidates who equate efforts to reduce global warming with impediments to economic growth. "We can't afford it" is a rhetorical phrase with no validity on this issue.

• Candidates who oppose raising auto fuel-efficiency standards probably won't support *any* attempts to fight global warming.

VOTE TO STOP OZONE DEPLETION

The U.S. is the largest manufacturer of ozone-depleting chemicals and accounts for 30% of their use worldwide.

I magine that it's the year 2050: Your great-grandchildren can only go outside to play if they wear special suits to guard against the sun's ultraviolet rays.

Could this really happen? No one knows.

But we do know that the ozone layer—a thin shield of gases in the upper atmosphere, which protects us from the sun's harmful ultraviolet rays—is disappearing. And there's no technology to replace it.

While efforts are being made—around the world and in individual cities and towns—to phase out ozone-depleting chemicals, people are not getting results fast enough. That's why we need to elect candidates at every level who are committed to saving the ozone layer…and doing it *now*.

DID YOU KNOW

• The ozone layer is being destroyed by man-made chemicals.

• The most commonly used of these are called chlorofluorocarbons (CFCs). Americans are the world's heaviest users of CFCs.

• CFCs are mainly used as refrigerants in home and car air conditioners, and refrigerators. They're also in foam insulation, solvents, aerosol dust removers, and other products.

• A single CFC molecule released into the atmosphere can last 100 years and destroy 100,000 ozone molecules.

• Other ozone-depleting chemicals include:

√ Halons (used in fire extinguishers). These are produced in smaller quantities than CFCs but are as much as 16 times more damaging to the ozone layer than CFCs.

√ Carbon tetrachloride (used in dry cleaning).

√ Methyl chloroform (a solvent used in industry and some consumer products, like aerosols). This is weaker than CFCs, but is produced in greater quantities.

√ HCFCs (used in polystyrene foam and other products). These are slightly altered chlorofluorocarbons that are less damaging—but are still ozone-depleting.

√ Methyl bromide (a pesticide used in fumigation).

Into Thin Air

• According to scientists at NASA, recent data shows that ozone-depleting chemicals are now eating away the ozone layer twice as fast as previously anticipated. And in 1992, for the first time, significant thinning of the ozone layer over North America was predicted.

• The EPA predicts that if nothing is done to restrict ozone-depleting chemicals, over the next 50 years 12 million Americans will develop skin cancer, and more than 200,000 of them will die from it.

• Thus far, the effects of ozone depletion are most severe in Australia and Antarctica. For example, in Australia, scientists say ozone depletion has already caused lower crop yields and a 300% increase in skin cancer.

• Ozone depletion could ultimately jeopardize all life on Earth, because stronger ultraviolet rays may destroy the microscopic plankton in the ocean that are the first links of the food chain. There is evidence that this is already happening in the Antarctic Ocean.

WHAT OUR POLITICIANS ARE DOING: *A few examples*

The White House. On February 11, 1992, President Bush announced that the U.S. would phase out CFCs, halons, carbon tetrachloride, and other ozone-depleting chemicals by the end of 1995. The plan, however, gave exemptions to undefined "essential uses." He also said he was reexamining U.S. policy on HCFCs and methyl bromide, but provided no other details.

Congress. In 1990, Congress passed amendments to the Clean Air Act requiring recycling of CFCs used in refrigerators and home and car air conditioners. (This will reduce ozone depletion by keeping CFCs out of the air.) However, the EPA (whose director was appointed by President Bush) has still not issued any regulations, the first step toward enforcing the law.

State and Local Governments. Many cities and states have enacted their own legislation to phase out CFCs sooner than required by the federal government. In 1989, for example, Irvine, California, became the first city to outlaw CFCs. Since then, at least 15 states and 25 cities have banned some or all of their uses.

• Some cities—including Denver, Colorado, and Newark, New Jersey—have passed laws requiring car air conditioner and refrigerator repair facilities to recycle CFCs.

• State action: Hawaii started mandatory recycling of car air conditioner CFCs in 1991. In Maryland, auto service stations that buy CFC recycling machines are allowed a 100% tax credit for them. And Vermont passed a law preventing new cars from being sold with CFC air conditioners.

THE POLITICAL DEBATE: *Some arguments you might hear*

Argument: "There's not enough proof that ozone depletion is really happening."

Answer: Years ago that argument might have made sense. But now the evidence is—unfortunately—overwhelming.

Some background: In 1974, when two scientists at the University of California suggested that CFCs might be destroying the ozone layer, many scientists scoffed…and most industry representatives called the theory alarmist. Even after CFCs were banned from aerosols by Congress in the late '70s, there was a concerted effort by industry representatives to discredit the ozone-depletion theory.

However, since then, satellite photographs and other research have confirmed not only that ozone depletion is occurring, but that it's occurring much faster than anyone anticipated.

Argument: "We're moving fast enough to save the ozone layer."

Answer: There's *no way* we can move fast enough. Even if we eliminated all ozone-depleting chemicals tomorrow, ozone depletion would continue for 50 to 100 years. The reason: It takes

CFCs 6 to 12 years simply to reach the ozone layer...and their destructive power may last as long as a century. And even after we stop manufacturing them, we'll still have to tackle the problem of disposing of the existing chemicals that haven't been released into the air yet, such as those contained in old air conditioners.

Even the Montreal Protocol, an international agreement that will phase out CFCs in industrialized countries by the year 2000 and in developing countries by 2010, didn't move fast enough. A number of countries have decided to accelerate their phase-out dates. Last year, for example, the 12-nation European Community pledged to stop using CFCs by 1997, while Germany passed a law banning CFCs by 1995.

Argument: "We need to produce some CFCs for 'essential uses' and to service existing equipment."

Answer: With recently developed recycling and cleaning technologies, there is almost no need for new or "virgin" CFCs. Today, for example, recycled CFCs are being widely used to recharge car air conditioners. And experts expect the recycling and reuse of ozone-depleting chemicals to become a big and **very** profitable business in the next few years.

In addition, businesses are finding substitutes for CFCs all the time. Five years ago, the computer industry thought there were no alternatives to CFC solvents for cleaning circuit boards...Today they're using soap and water. And Hughes Aircraft replaced its CFCs with simple, inexpensive chemicals derived from lemon juice.

A rigorously enforced phase-out of the chemicals will encourage businesses to focus their energies on coming up with more safe substitutes.

VOTE FOR THE EARTH: *Picking the right candidates*

√ Do candidates understand the urgency of the problem? According to Jeff Tryens of the Center for Policy Alternatives, "Many people running for office know very little about ozone depletion and think the problem has already been solved."

√ If all candidates seem to understand the problem, see who's willing to act fastest to solve it.

VOTE FOR THE RAINFORESTS

Over the past decade, the world has been losing rainforests at the rate of about 42 million acres—an area the size of Washington State—every year.

The Earth's tropical rainforests are vital to our survival. They fight global warming, prevent floods and drought, provide a home to millions of species of plants and animals, and are even the source of about 25% of the medicine we rely on. Many medical researchers feel that if we find a cure for AIDS or cancer, it's likely to come from as-yet-undiscovered plants or animals in the rainforests.

Yet, *every minute* of the day, 80 *acres* of rainforest are destroyed—burned to make space for cattle or clear-cut for timber. According to The Rainforest Action Network, at this rate of deforestation, all of the Earth's rainforests will be gone in as little as 25 years.

This problem has gotten so much publicity that you might think our politicians are doing everything possible to solve it. But they're not. Through organizations like the World Bank, we're actually funding projects that *cause* rainforest destruction. And there are virtually no restrictions on importing rainforest timber into the U.S.

In fact, some politicians say we "can't afford" to help save the rainforests. But for our children's sake, we need to elect people who know we can't afford *not* to.

RAINFOREST FACTS

- About 50% of the world's rainforests are already gone...and we're losing tropical forest nearly twice as fast today as we were ten years ago.
- 80% of all Amazonian deforestation has taken place since 1980.
- Between 1960 and 1985, 40% of all Central American rainforests were cleared to create pasture for beef cattle.
- Severely damaged rainforests may never recover—once they're

gone, they're gone for good. The soil in a rainforest is not rich—only about a 2-inch layer contains any nutrients. Most of a rainforest's nutrients are stored in the vegetation.

• Although rainforests make up only 6% of the Earth's surface, they house at least 50% of the world's species. Experts estimate that clearing tropical forest may drive as many as 50,000 species a year into extinction.

• About a third of the carbon dioxide—the main "greenhouse gas" released into our atmosphere—comes from burning rainforests.

WHAT OUR POLITICIANS ARE DOING: *A few examples*

Congress. Through the World Bank, Congress funds large development projects in Third World countries. Many of these projects contribute to rainforest destruction—so our tax dollars are currently helping destroy rainforests. If more members of Congress were rainforest activists, the same money might be used to fund sustainable projects—projects that help Third World countries develop... *without* destroying their resources.

• About 25% of rainforest destruction is due to the timber trade. To help U.S. consumers identify rainforest wood—and avoid buying it—members of Congress proposed the Tropical Forest Consumer Information and Protection Act. It would require the labeling of all imported tropical wood.

The Treasury Department (Executive Branch). In 1990, the Treasury Department changed the terms on a small portion of the debts that developing nations (where most rainforests are located) owe the U.S. The good news: New terms allowed them to pay back only the principal and put the interest in an environmental trust fund for themselves. The bad news: Before countries could qualify for the terms, they had to meet other economic criteria...which actually promoted environmentally destructive development.

State and Local Governments. More than 15 state and city governments have passed laws prohibiting the purchase of tropical timber for government-funded construction projects, or furniture made with tropical wood. A few cities are even considering an outright ban on tropical timber.

THE POLITICAL DEBATE: *Some arguments you might hear*

Argument: "Other countries have a right to make money with their own resources any way they choose."

Answer: Yes, but their policies not only affect global health, they're bad economics. Third World countries clear-cut their rainforests for agriculture, logging, cattle ranching, population resettlement, etc. But once the trees are gone, rainforest soil becomes arid and infertile. It can turn to desert in as little as three years. Then not only are the trees gone, but the land becomes worthless for subsistence—and it's subject to flooding and drought as well.

On the other hand, there are things the countries can do that will support their economies indefinitely...and save the forests at the same time. Research shows that an acre of rainforest that's sustainably harvested for fruit and rubber products can be worth as much as 12 times more than when the land is cleared. And the potential value of its plants for pharmaceuticals is enormous.

Argument: "We shouldn't be meddling in other countries' affairs."

Answer: The U.S. isn't trying to infringe on anyone's sovereignty. But we have a right to insist that our government use our tax dollars to support projects that protect the rainforests. Without "meddling," the U.S. can:

√ Prohibit the World Bank and other organizations from using U.S. contributions to fund mining projects, hydroelectric dams, and other projects that destroy rainforests.

√ Revoke favorable trade status for products that have been produced through rainforest destruction. The U.S. can also place higher tariffs on them...or ban them altogether.

√ Inform governments of rainforest countries that if they want loans, grants, etc., they must meet certain conditions.

√ Forgive some foreign debt in exchange for commitments to protect forest area. We've used this approach before—for example, we forgave $7 billion worth of Egyptian debt when they supported the U.S. in the Iraq War. Rainforest protection is at least as important to our national interest.

Argument: "Rainforest destruction is not our problem."

Answer: This is a global problem. It affects all of us—not only in obvious ways, like adding to global warming and threatening our source of life-saving drugs, but in ways we may not have thought about. For example: If the rainforests disappear, the land will be barren; the millions of people who live there will starve. So if we don't help them learn to use their forests in a sustainable way now, we'll have to deal with a vast influx of refugees or help them with disaster relief later.

Another example: The measurable decrease in the number of songbirds in the U.S. is directly attributable to the destruction of rainforests in Central and South America, where the birds spend much of the year. Besides being beautiful, these birds help control insects. If they're wiped out, some insect populations could explode.

VOTE FOR THE EARTH: *Picking the right candidates*

√ Do they talk about the rainforests as a critical issue?

√ Do they understand that what happens to the rainforests affects us and our children?

√ Do they support legislation to restrict U.S. imports of tropical timber?

√ Are they willing to ban the use of tropical timber in local or national government-funded projects?

√ Are they willing to ban government purchase (local or national) of furniture made with tropical timber?

Voter Notes

• Be wary if a candidate is supported by any member of the International Hardwood Products Association.

• Another way to tell if candidates will support rainforest protection: check out the way they approach other forest protection issues.

VOTE TO SAVE OUR OCEANS

Ships and oil-drilling platforms dump 14 billion pounds of garbage into the seas each year—the equivalent of more than 1.5 million pounds every hour.

Our oceans are in trouble.

Right now, every ocean on the planet is polluted...and marine life is dying out.

This doesn't just mean a loss of recreational beaches and sport fishing. The oceans keep us alive: they're an essential source of oxygen and food, and they affect global weather patterns...yet we treat them as if they're expendable.

The people who decide how every issue that affects the oceans—from how many fish can be caught in our waters to whether pollution should be dumped into them—are our elected officials.

That's why it's so important to support candidates who are committed to protecting our oceans.

TRASHING THE OCEAN

• The world's shipping industry dumps more than 450,000 plastic containers into the seas every day.

• U.S. industries and cities dump an estimated 6.4 trillion gallons of waste and sewage into the oceans every year.

• From 1980 to 1989, oil spills in U.S. waters averaged 3 million gallons a year.

• On any given day, a third of U.S. shellfish beds are closed due to pollution.

• The National Academy of Science estimates that every year, the U.S. is the source of a third of the "total world marine litter" dumped in the ocean.

• Fish populations are being depleted at a staggering rate. Every night, fleets on the high seas fish out an area of ocean equivalent to the size of Ohio. This has led to fish shortages. For example, a recent study by the National Marine Fisheries Service revealed that of 153 fish species in U.S. waters, 42% have fallen *below* sustainable levels.

WHAT OUR POLITICIANS ARE DOING: A *few examples*

The White House. President Bush has declared a ten-year moratorium on offshore oil drilling in California and Florida. But this moratorium is temporary; its future after the year 2000 is questionable. And the president has steadfastly refused to ban oil drilling in numerous sensitive marine and coastal areas—particularly in Alaska.

Congress. There are many opportunities for members of Congress to protect the ocean:

• The National Marine Sanctuaries Act, which protects vital marine habitats (like the coral reefs of Florida and the submarine canyons of Monterey Bay, California) from pollution and development schemes, is up for reauthorization in 1992. Less than 1% of U.S. waters are protected by legislation, and the ten marine sanctuaries currently in existence are as much a part of our heritage as Yellowstone or Yosemite national parks.

• The Magnuson Fishery Conservation Management Act, which sets U.S. fishing quotas, comes up for reauthorization in 1993. Currently, these quotas are set by government officials and representatives of the fishing industry. Environmentalists see this as a case of the fox guarding the henhouse. The decision-making process should include marine scientists, the public, and environmental experts.

• The Marine Mammal Protection Act, which protects seals, dolphins, whales, otters, polar bears, and other marine mammals in U.S. waters, is up for reauthorization in 1993. The fishing industry is resisting restrictions on fishing at the expense of our marine life.

State and Local Governments. Local governments have an impact on the ocean, too. They can initiate actions that alter or modify federal actions. For example: In Florida, federal interest in leasing offshore oil tracts was successfully fought by local and state governments.

• In fact, local efforts (often with federal support) have been among the most successful in preserving and restoring threatened marine areas. For example: Cleanup efforts in the Potomac and Delaware river estuaries have restored fish populations (bass and shad) that were almost wiped out by pollution.

THE POLITICAL DEBATE: *Some arguments you might hear*

Argument: "Protecting jobs is more important than protecting the ocean."

Answer: A healthy ocean provides millions of jobs around the world, so protecting the ocean *is* actually protecting jobs.

However, even when it appears to some people that we have to make a choice between jobs and marine life, it's not necessarily true. For example: For 10 years, marine conservationists have been fighting a battle to protect sea turtles that get caught in shrimp nets. (This is the number-one cause of death of sea turtles at human hands, according to the National Academy of Sciences.) The solution proposed was a "turtle excluder device," which allows turtles to escape from nets without disturbing the catch.

Shrimpers insisted that using this device was too costly—it would destroy the industry and put people out of work. But the excluder device is now mandated by federal law...it hasn't cost the industry jobs...and thousands of turtles have been saved.

Argument: "What about fishing quotas? *They* cost jobs."

Answer: While overfishing ensures jobs now, it will cause massive job losses in the future, when most fish are gone. Already, for example, overfishing by shrimp trawlers—which catch 10 pounds of other fish for every 1 pound of shrimp—is affecting fish stocks and jobs in the red snapper industry. And in the North Atlantic, the number of breeders in the bluefin tuna population (a popular sport and commercial fish) has fallen from 250,000 in 1970 to 22,000.

Argument: "It's safe to dump garbage and contaminated waste in the ocean if it's done far enough from land. The water dilutes the material, or it sinks harmlessly to the bottom."

Answer: The idea that the oceans are so big that they can somehow always cleanse themselves is a dangerous myth. Once dumped, pollution remains in the ocean's complex circulation system. Debris on the ocean floor can be stirred up by currents and fish movements years after it has settled. And as with other forms of pollution, we don't know what (if any) contamination levels are safe in the oceans. Scientists believe pollution impairs the immune systems of fish and marine mammals, leaving them more vulnerable to infection and disease. In the 1980s, for example, Europe's harbor seal population was cut in half by mass die-offs tied to a viral infection. More recently, a fish called the Atlantic menhaden—a mainstay of the U.S. fishing industry—has been attacked by a fungus that literally eats holes in the fish's body.

These animals are sending us a message that poisons are beginning to overwhelm their systems—and since we depend on marine life for food, the poisons are getting into our systems as well.

Argument: "Offshore oil drilling may be an environmental risk, but it is essential to safeguard our energy security."

Answer: No matter how you look at it, offshore oil drilling isn't worth the risk. First of all, there's not that much oil off U.S. shores: Excluding the Gulf of Mexico, it's estimated that there are 4.7 billion barrels of recoverable oil under federal waters. This might seem like a lot, but the U.S. consumes 18 million barrels of oil a day—so the 4.7 billion barrels wouldn't even last a year. According to the Center for Marine Conservation, there are only a few weeks' worth of oil off the coast of California and a few days' worth in the Florida Keys.

Second, offshore oil drilling is very expensive—it costs 2 to 3 times more than land-based drilling.

And finally, drilling for offshore oil is hazardous to marine life and nearby shore areas.

VOTE FOR THE EARTH: *Picking the right candidates*

√ Do they support reauthorization of the National Marine Sanctuaries Act and a strong Marine Mammal Protection Act?

√ Do they support a permanent moratorium on oil drilling and transportation in sensitive marine areas?

√ Do they believe the Magnuson Act should be amended so fishing quotas are set by marine scientists, not the fishing industry?

VOTE FOR FAMILY PLANNING

*During the 1990s, the equivalent of the current population
of China will be added to our planet.*

The size of the human population is linked to virtually every
environmental condition on Earth—from traffic jams to
acid rain, from deforestation and global warming to food
and water shortages.

That's probably why, in a recent national poll, an overwhelming
majority of the people surveyed said that overpopulation is the
biggest environmental problem we face today.

World population now stands at about 5.4 billion people. Scientists estimate that at the current rate of growth, and without deci-
sive action, the population will double by the year 2040...and keep
growing.

Can the Earth support a population this size?

No one is sure...which is why we need to elect people who will
put the enormous resources of government behind family planning
and other programs to improve the status of families—*now*.

GROWING, GROWING...

• The world population has more than doubled since 1950.

• In the time it takes you to inhale and exhale, three more babies
will be born.

• On average, 16,000 people are born every hour, 90% of them in
developing countries.

• There are currently more women of childbearing age in America
than ever before, and the rate of unintended pregnancies in the
U.S. has reached an incredible 50%.

FAMILY PLANNING

• Studies show that, currently, there are an estimated 125 million
couples worldwide who want to limit their family size...but don't
have access to modern contraceptives.

- During the 1990s, the number of reproductive-age couples in developing countries will grow by about 18 million per year—and most will want access to modern contraceptives.

- The United Nations estimates that making family planning services available to everyone who wants them will cost about $10 billion a year (the equivalent of about 3% of the annual U.S. military budget).

- But in 1990, the U.S. and all other donor nations *combined* spent only $754 million on family planning for developing countries. That's only about two-thirds of the amount Americans spend on *birdseed* every year.

- Due to high product liability costs and a hostile regulatory system, the number of U.S. companies involved in contraceptive research has dropped from nine to one in the last 15 years.

WHAT OUR POLITICIANS ARE DOING: *A few examples*

The White House. Under presidents Reagan and Bush, funding in real dollars for family planning programs has not increased in the past 12 years. And new policies restrict how the money the U.S. *does* spend can be used.

- For example: In 1985, the Reagan administration cut off all U.S. funding for the United Nations Population Fund (UNPFA). The administration alleged that the U.N. agency participates in coercive family planning programs in China. This is a legitimate concern, but a study done by the Agency for International Development (a U.S. government agency) found no evidence that the UNPFA "supports or participates in any of the coercive activities alleged to take place in China." The principal effect of the Reagan administration decision was to punish the other 140 countries to which UNPFA provides desperately needed family planning, maternal, and child health-care services.

Congress. Congress effectively defended the international population assistance program from significant cuts launched by the Reagan administration. But it hasn't aggressively increased funding levels to meet requests by developing nations.

- In 1992, Congress approved up to $250 million in family planning…but it should have appropriated $570 million to meet the U.S.'s share of the U.N. drive to provide worldwide access to family planning.

• Title X of the Public Health Service Act, the only federal program devoted to family planning, has lost funding steadily in the past decade. In 1980, the budget was $162 million. In 1992, it will be $150 million. Adjusting for inflation, that means spending power has dropped by two-thirds.

THE POLITICAL DEBATE: *Some arguments you might hear*

Argument: "Overpopulation isn't America's problem; our birthrate is lower than most countries'."

Answer: When it comes to global problems like ozone depletion, deforestation, the greenhouse effect, and overpopulation, national boundaries are meaningless. More people means more pollution, more garbage, more energy use, etc. It puts all of us—and all of our children—in jeopardy.

Argument: "Birth control is too personal and controversial for government to be involved with."

Answer: This issue involves the health and well-being of every citizen. Regardless of controversy, governments can't ignore the need for birth control. In fact, it's their *responsibility* to do something about it. Already, three million women and children around the world die every year from lack of family planning. (The reason: when children are born closer than 24 months apart in the Third World, their survival rate goes down 50%.)

Argument: "No one should tell someone else how many children to have."

Answer: We agree. That's why we think birth control should be available to people who want it.

Stabilizing the population now with *voluntary* measures—instead of waiting until it reaches crisis proportions, as in China—is the best way to prevent mandatory family planning from ever happening.

Argument: "Technology is the answer, not family planning. The planet can support a lot more people if we develop cleaner, more efficient ways of living."

Answer: Most scientists believe that while technology can lighten the burden of population growth, it isn't a solution. In fact, two prestigious scientific organizations—the U.S. National Academy of Sciences and the Royal Society of London—recently released a first-ever joint statement saying that the present world population growth of nearly 100 million people annually, in conjunction with overconsumption of resources, is the principal cause of deforestation, global warming, and the "unparalleled" pace of species extinction. In their opinion, no possible amount of pollution control, energy efficiency, or other form of conservation can keep up with the problems caused by overpopulation. They specifically recommend *more rapid* world population stabilization, beginning with family planning and improving the status of women and children.

VOTE FOR THE EARTH: *Picking the right candidates*

√ Do they consider stabilizing population levels a positive goal?

√ Do they support a "national population policy"? This would provide a forum for discussing the impact of population growth on all issues—from unemployment to the environment.

√ Are local candidates willing to support tough land-use and family planning policies to help create a permanent, sustainable and stable community?

Voter Notes

• National health care is a major issue this year. Check to see whether candidates believe that a public health insurance plan should include coverage for family planning services.

• Watch out for candidates who avoid discussing family planning and overpopulation altogether.

• Watch out for candidates who support more growth without putting a clear priority on environmental protection.

SAVING

AMERICA

VOTE TO CONTROL TOXIC WASTE

Every day, U.S. industries produce the equivalent of nearly 30 pounds of toxic waste for every man, woman, and child in the country.

American businesses produce more than a billion tons of hazardous waste every year—from industrial by-products to dry-cleaning chemicals, from nuclear fuel to radioactive materials used in medical treatments—and no one's sure what to do with it. As a result, toxic waste is often dumped in places where it can pollute water, soil, and the air. Often, it is stored dangerously close to people's homes.

That means your community, your neighbors, and your own family may be at risk.

There's no way any individual can single-handedly stop companies from creating hazardous waste. But the people we elect can create and enforce regulations that keep toxic waste under control.

DID YOU KNOW

• There are 70,000 industrial chemicals used in the U.S.—and complete health studies have been done on only 1,600 of them. Any of these chemicals could be dumped in your area.

• There are at least 2,000 serious accidents involving hazardous waste in the U.S. each year.

• The EPA has identified 32,000 potentially toxic waste sites in the U.S. More than 1,000 of them are already considered "an immediate health threat" by the U.S. government...but in the last decade, the EPA cleaned up only 89 of the most contaminated sites.

• The EPA estimates that 99% of all Americans already have some level of toxic compounds in their bodies.

WHAT OUR POLITICIANS ARE DOING: *A few examples*

Congress. Companies weren't telling people whether they were discharging toxic chemicals in their neighborhoods. So, in 1986,

Congress passed the Community Right to Know Act, which requires companies to report on emission of 328 key chemicals. However, the law isn't strong enough: it contains so many exemptions that only 1 out of every 20 pounds of toxic emissions winds up being reported.

• Now a new law, the Community Right to Know *More* Act, is being debated in Congress. It would eliminate many exemptions and require companies to develop plans to reduce toxic chemical use.

The EPA (Executive Branch). In 1980, Congress passed a law known as Superfund, which is supposed to provide the EPA with the money needed to identify and clean up toxic waste sites. But although more than $7 billion of Superfund money has been spent, the EPA has succeeded in cleaning up fewer than 1% of the toxic waste sites in the U.S.

• One reason: Companies find it more cost-effective to fight the EPA in court than to clean up their toxic waste. As a result, 70% of the Superfund money—almost $5 billion—has gone to legal costs.

The Nuclear Regulatory Commission (Executive Branch). The Commission has established a new category of nuclear waste called "Below Regulatory Concern" (BRC). Even though BRC material is carcinogenic, nuclear plants can now dispose of it anywhere they choose—including regular landfills and sewers. They can also sell the materials for use in consumer products, such as frying pans, zippers, and belt buckles. So you could be *wearing* radioactive, cancer-causing toxic waste and not even know it.

• Twelve states and more than 60 cities have already passed laws to control the disposal of BRC waste. However, the NRC has announced that it will allow waste generators to dump waste in these states, anyway.

State Governments. Sixteen states have passed "pollution-prevention" laws that help companies reduce the use of toxics by providing technical assistance, grants, and education, while requiring them to set reduction goals and report their progress to the public.

THE POLITICAL DEBATE: *Some arguments you might hear*
Argument: "We don't need to regulate toxic waste—industry will

take care of the problem on its own."

Answer: Don't bet on it. A few companies *have* voluntarily reduced toxic waste…but they're the exception, not the rule. A 1986 government study found that U.S. industries could cut the amount of hazardous materials they produced in half in only five years—using *existing technologies*. But by 1992—six years later—very few companies had done it.

Argument: "Cutting down on the use of toxics isn't cost-effective for business."

Answer: First of all, according to *The Wall Street Journal*, reducing reliance on toxic chemicals can actually *save* companies money. A case in point: A Du Pont Company plant in Beaumont, Texas, reduced its use of toxic materials, cut its pollution, and now saves $1 million a year in related costs—like hazardous chemical processing and cleanup—not to mention what it's saved in lawsuits and fines.

And second, if the American public knew how much the use of toxics already costs us—in tax dollars, medical costs, and health problems—we'd have a law restricting their use right away. Some experts estimate that cleaning up our toxic waste sites could cost more than *$1 trillion*. If we don't take care of this problem, our children will inherit it.

VOTE FOR THE EARTH: *Picking the right candidates*

√ Do they support the Community Right to Know More Act?

√ Look for candidates who support toxins reduction programs. Massachusetts and New Jersey have pioneered programs that require an actual reduction in the use of toxic chemicals. It's time for other states to follow suit.

√ Do local candidates oppose the construction of toxic waste disposal facilities in residential neighborhoods? This is especially critical in low-income areas, where companies typically try to locate such facilities.

Voter Notes

• If a candidate gets large contributions from chemical-industry groups, like the American Petroleum Institute or the Chemical Manufacturers' Association, watch out.

VOTE FOR WETLANDS

*In the last 200 years, the U.S. has lost an average
of about one acre of wetlands every minute.*

Until the 1950s, wetlands were generally considered a waste of space. They were "worthless" swamps that stood in the way of agriculture, factories, houses, and shopping malls. So people filled them in. By 1990, more than half of America's wetlands were gone.

Now we realize that they're a critical part of a healthy environment. Wetlands are natural filters that fight pollution in our waterways; they're essential breeding grounds for fish, birds, and other aquatic life; they provide natural flood control; they provide food for plants and animals. In fact, without wetlands, life as we know it could not be supported.

DID YOU KNOW

• Wetlands are generally the boundary area between solid land and a river, stream, or ocean. They are nature's sponges.

• Wetlands include bogs, swamps, salt marshes, and prairie potholes. But they don't have to be wet all year to play an important ecological role. The most common type of wetland in the U.S.— over 60% of the total—is forest and shrubland (seasonal wetlands).

• Two hundred years ago there were 221 million acres of wetlands in the continental U.S. Fewer than half remain.

• Each year we lose an average of 300,000 acres of wetlands.

• Destroying wetlands eliminates soil and vegetation that absorbs water and prevents floods. When wetlands are gone, taxpayers end up spending billions of dollars for flood-control structures.

• Wetlands are the nurturing grounds for an abundance of all aquatic species. It has been estimated that 1/3 of the U.S.'s endangered species depend on wetlands for food and habitat.

WHAT OUR POLITICIANS ARE DOING: *A few examples*

The White House. In 1988, President Bush pledged "no net loss" of wetlands. But, in 1991, he tried to slip around his promise by proposing new guidelines that would dramatically change the federal definition of wetlands.

• According to the federal government's own scientists, the president's proposal could result in the loss of *almost half* of our remaining wetlands—including 25% of the Everglades.

• Public protest has stopped the move for now. But the White House hasn't given up. In May 1992, the Bush administration drafted another proposal that would allow farmers to drain up to ten million acres of wetlands.

Congress. The Clean Water Act (which regulates wetlands) will be up for reauthorization in 1992. So far, two very different amendments to the law—one good, one bad—have been introduced:

√ HR 4255 (the Edwards Bill), which environmentalists support, calls for the term "wetlands" to be defined by scientists, not politicians. It would also tighten controls on draining and developing wetlands.

√ HR 1330 (the Hayes Bill) is even worse than Bush's proposals. Activists say it could cost us up to 80% of our remaining wetlands.

• The bad news: There's currently more support for the Hayes Bill.

State and Local Governments: Many local governments protect wetlands by restricting development. This is often the only way to save them.

• However, because some of these policies have been so effective, developers are trying to undermine them. In some states—for example, Washington—bills have been introduced that would take the power to protect wetlands away from communities and give it exclusively to state government.

THE POLITICAL DEBATE: *Some arguments you might hear*

Argument: "To qualify as a wetland, land should be wet all the time."

Answer: That's a common misconception...and a dangerous one, because developers use it to justify eliminating wetlands.

The term "wetlands" doesn't mean "wet on the surface all the time." It refers to a unique combination of soil, plant life, and water conditions that must be present:

√ Water-loving plants

√ Soils that develop under the influence of water

√ The presence of water either on the surface or saturating the soil below the surface, where the plants' roots are

Some wetlands are never wet on the surface, and some are only under water at certain times of the year. Others are wet on the surface only when flooding takes place, like flood plain areas next to riverbanks. Whether they're wet on the surface all year-round, or a few days a year like the prairie potholes in the Midwest, they are equally important to our survival.

Argument: "We have to keep filling in wetlands if we want to keep our economy growing."

Answer: Nonsense. Wetlands are already extremely valuable...as wetlands. And losing them costs us billions of dollars. For example:

√ Fish that feed and spawn in wetlands account for 71%—or $14 billion—of the U.S.'s annual commercial catch. Even loss of nearby wetlands can hurt fish by damaging the water quality and their food supply.

√ As wetlands disappear, the seafood industry suffers. (Shrimp and crabs, for example, breed in wetlands.) There's already been a sharp decline in seafood harvesting in Southeastern coastal areas.

√ Wetlands naturally absorb overflow from rivers and watersheds. The U.S. has spent more than $7 billion on flood control in the last 20 years. That amount will increase as wetlands disappear. For example: In 1987, following the destruction of local wetlands, a flood in DuPage County, Illinois, caused $120 million in damage; the estimated cost of the county's new flood-control project is $100 million more.

√ According to the Environmental Defense Fund, the cost of improving sewage treatment to make up for the wetlands that could be lost because of the Bush proposal would be $70 billion.

√ Sport hunting and fishing—a $28 to $40 billion industry—is being threatened. For instance, wetland losses in the last 35 years explain much of why the duck population has dropped by 20%.

Argument: "Wetlands protection interferes with private property rights. The government is 'taking' private property."

Answer: Anyone who buys property in America knows from the start that there are going to be legal restrictions on how the land can be used. It's the government's responsibility to impose restrictions that would prevent property owners from harming their neighbors...and the public. For example: When wetlands are filled in, they often cause flooding on neighbors' property downstream.

But the truth is, wetlands regulation rarely prevents a property owner from building *somewhere* on their property. Most people who apply for permits to build on wetlands get them—with some restrictions intended to minimize or avoid harm to the land.

VOTE FOR THE EARTH: *Picking the right candidates*

√ Who do candidates think should define what a wetland is: scientists or politicians? Pro-Earth candidates say scientists should.

√ Which amendment to the Clean Water Act do national candidates support—the Hayes Bill or the Edwards Bill? (Environmentalists support the Edwards Bill, HR 4255.)

√ How do local candidates feel about land development issues like shopping centers, office buildings, etc.? Are they for careful "planning and management"...or are they staunchly pro-growth?

Voter Notes

• Watch out for candidates who talk about the need for wetlands policies that "balance economic and environmental interests." That's usually a smokescreen for a "pro-growth" position.

• "No net loss" is a positive term that means we won't lose any net acreage of wetlands. If a candidate really means this, he or she is in favor of stopping all activities that destroy wetlands. Unfortunately, President Bush—who used the term in his 1988 campaign—showed us that it can also be meaningless.

VOTE TO SAVE ENDANGERED SPECIES

Experts estimate that as many as 50,000 plant and animal species become extinct every year.

According to the National Wildlife Federation, species are becoming extinct faster today than at any time since the disappearance of the dinosaurs 65 million years ago. Think about that.

Under normal circumstances, 1 to 10 known species become extinct every year. That's a part of natural selection. But now we may be losing 100 species *every day*. Two-thirds of the world's monkeys and apes are threatened with extinction. Nearly half of the world's turtles are endangered. Three-quarters of the world's bird species are either losing numbers or are endangered.

What are we doing to stop this frightening disappearing act? America's strongest defense against it is the Endangered Species Act, first passed in 1973. Now it's up for reauthorization—which means the people *we* elect are going to be deciding which animals will survive, and which will be lost forever.

THE ENDANGERED SPECIES ACT

How It's Supposed to Work

• Anyone can petition the Fish and Wildlife Service (FWS) to put a species on the endangered species list. The FWS is then required to determine if there's enough evidence to support a listing.

• That review takes about a year. Then, if the agency formally proposes an "inclusion," it takes another year for the species to get listed.

• Once a species is put on the list, the Department of the Interior (DOI) has to present a "recovery plan"—a blueprint detailing how it intends to bring the species back to the point where it's no longer threatened or endangered.

How It Really Works

• Unfortunately, since the Endangered Species Act (ESA) was first passed, the law has been consistently underfunded. In 1991, the entire ESA budget was $55 million—the same amount Americans spend betting on dog races in one week.

• The backlog of plants and animals waiting to be considered for protection under the ESA is estimated at 3,500 species. "Of those," says the National Wildlife Federation, "about 1,000 are now considered 'Category 1'—species that the Fish and Wildlife Service recognizes *should* be listed for scientific reasons, but aren't, due to underfunding."

• Activists say that it has taken more than 10 years to get some recovery plans through the Department of the Interior. To date, about 50% of listed species don't have recovery plans—which means little is being done to save them.

WHAT OUR POLITICIANS ARE DOING: *A few examples*

The White House. The president wants to weaken the Endangered Species Act, using his appointees at the Department of the Interior.

• The department's secretary, Manuel Lujan, has been an active opponent of the ESA. For example, his department's Fish and Wildlife Services ruled that the Louisiana black bear was "threatened," not "endangered"—even though there are only about 100 in existence—so that logging in southern forests could continue. And a Department of Interior scientist has testified that she was told by superiors to alter her scientific opinion to allow development that destroys habitat essential to the survival of the Mt. Graham red squirrel.

• Secretary Lujan has the authority to convene a special committee known as the "God Squad." It has life-and-death power over endangered species because it can override the ESA in favor of business and allow actions that clearly put a species at risk.

Congress. The ESA is under attack in the House, where radical anti-environmentalist forces have introduced at least 3 different bills aimed at weakening the law. One, H.R. 4058, would require a cost-benefit analysis for each endangered species. That means the survival of every animal and plant would be determined by its effect on business profits.

• Another bill, H.R. 5105, would allow the Secretary of the Interior to grant businesses special exemptions to the ESA. Again, it would give the government the authority to arbitrarily protect a company's profits instead of a species.

• The good news is that Rep. Gerry Studds of Massachusetts has introduced a package of amendments—H.R. 4045—that would strengthen the ESA and almost double its funding.

• No action is expected on reauthorization until after the November elections. So it's especially important to elect environmentally concerned candidates to Congress this year.

State Governments. Most states have their own endangered species legislation. Generally, they either adopt the federal act verbatim, or adapt it and include their own listings of native plants and animals. Legally, state statutes must be at least as restrictive as the federal act. But many are more ambitious and do a better job of protection.

Local Governments. Local efforts have saved some endangered species. For example, in 1985, Brevard County, Florida, passed a special ordinance to protect local beaches, where female sea turtles lay their eggs. Since then 12 counties and 23 cities in Florida have passed similar laws, and scientists are hopeful that survival rates for turtle hatchlings, and ultimately adult turtles, will increase.

• The city council of Ft. Collins, Colorado, made their city an urban wildlife sanctuary to protect plants, animals, and wetlands by educating the community and protecting land from development.

THE POLITICAL DEBATE: *Some arguments you might hear*
Argument: "Animals aren't as important as people."
Answer: People who say this don't understand how our ecosystem works. Every living thing on the planet is connected.

It's important to remember that animals serve as indicators of the health of the planet. Decades ago, coal miners used to take canaries with them into mine shafts as a sort of "early warning" system. If toxic gases were released into the air, the canaries, which are smaller and more sensitive to the gases than the miners, would die. The disappearance of huge numbers of animals is our warning system. Our life support system is in jeopardy.

Argument: "We don't have to save every single subspecies."

Answer: We cannot presume to know what role every species plays in maintaining life on the planet. And if we do guess, we could be wrong. We've been wrong before. For example, less than a decade ago, yew trees were considered little more than "large weeds" that got in the way of logging operations in the Pacific Northwest. Today, we know that they are the only source of taxol, a new drug that could save the lives of thousands of women with breast and ovarian cancer.

Scientists say the existence of subspecies provides plants and animals with the genetic variety and resilience they need to adapt to changing circumstances. So having subspecies is like having a genetic bank. We don't know what we're going to need—and it could come from any animal or plant. So we'd better save as many species as we can.

Argument: "The Endangered Species Act blocks economic development. We need to take economic and social considerations into account when deciding which plants and animals are endangered."

Answer: Of course these considerations matter—in fact, they're a major factor in implementation of the Act at virtually every stage. But that's a separate issue from the scientific decision about whether or not a species is endangered.

Obviously deciding whether a species is endangered is a scientist's domain, not a politician's or business person's.

Argument: "Environmentalists are using the ESA to stop activities they don't like—such as the logging of old-growth forests in the Pacific Northwest—and not just to save animals."

Answer: The explicit purpose of the law is to protect not only endangered species, but "the ecosystem upon which endangered and threatened species depend." That includes forests.

VOTE FOR THE EARTH: *Picking the right candidates*

√ Do they support strengthening or weakening the Endangered Species Act?

√ Do they favor increased funding to carry out the Endangered Species Act?

√ Which amendment to the Endangered Species Act do candidates support? Environmentalists favor H.R. 4045—the Studds amendment.

√ Find out if candidates support local efforts to protect wilderness areas essential to the survival of endangered species.

Voter Notes

• Beware of candidates who attack endangered-species regulations as the work of "environmental extremists." This is a smokescreen often used by timber, mining, and other business lobbyists to discredit strong laws to save endangered plants and animals.

• Watch out for candidates who talk about the need for "balancing" biodiversity with "human needs" or "economic considerations." They're likely to vote to weaken endangered-species laws.

VOTE FOR ENERGY EFFICIENCY

Americans spend more money on imported oil in a week than the U.S. government spends on energy efficiency research in a year.

I f all of our energy came from the wind or sun, energy wouldn't be such a big environmental issue. But the fact is, we get about 85% of our energy by burning fossil fuels—coal, oil, and natural gas. And the environmental consequences are devastating: smog, acid rain, global warming, water pollution, toxic waste, and more.

What can we do about it? Until we develop a cleaner energy system, the best solution is to save energy.

Most people don't immediately think of voting when they think of saving energy. But our elected officials decide everything from whether to use energy-saving light bulbs in government offices…to tax incentives for energy-efficient businesses…to our entire national energy strategy.

It's up to us to make sure these people are committed to saving energy—not wasting it.

DID YOU KNOW
• Although Americans make up only 5% of the world's population, we consume about one quarter of its energy.

• Americans spend $1 million on energy every minute.

• The U.S. spends about 10% of its GNP on energy purchases, while our major trade competitor, Japan, spends only about 5% of its GNP on energy. This "energy inefficiency tax" makes us less competitive in world markets.

• The federal government is the largest single user of energy in the U.S.

• Cars and light trucks emit about 1/3 of America's CO_2 (the major greenhouse gas), 1/3 of the nitrogen oxides (a cause of acid rain), and 1/4 of the hydrocarbons (which cause smog).

• Utilities that supply electricity and natural gas use a third of the nation's energy...and contribute about 50% of the CO2.

• The government has been less than enthusiastic in supporting alternative forms of energy. For example: It was reported that in 1991, the Navy proposed spending $75 million on solar equipment, explaining that it could save $150 million on fuel...*each year*. But so far, no action has been taken.

• In the 1980s, the U.S. was dominant in solar technology—controlling 70% of the market. Now we have only 30% of the market, and Japan has become the industry leader.

WHAT OUR POLITICIANS ARE DOING: *A few examples*

The White House. After the Kuwait crisis, President Bush introduced his "National Energy Strategy." It actually calls for Americans to use 37% more energy in 2010 than we're using today—including 12% more oil, 22% more natural gas, and 47% more coal (the most polluting fossil fuel of all). It also calls for doubling our dependence on nuclear power by 2030. Significantly, it does not call for improvements in automobile efficiency.

Congress. The National Energy Bill, passed by both the House and the Senate in 1992, will make America's electric and natural gas system more efficient, but will also remove public safeguards in the licensing of nuclear power plants. It offers some financial support for renewable energy, but gives much bigger subsidies to the oil industry and does nothing substantive to reduce U.S. oil use.

• Since 1975, Congress has fought continuously over fuel economy standards. A bill has been introduced in the Senate requiring a 40% improvement in motor vehicle energy efficiency by 2001. If Congress passes it, by the year 2005 we'll be saving an average of 100 million gallons of gasoline—and keeping 2 billion pounds of carbon dioxide out of the atmosphere—*every day*. But U.S. carmakers have fought new legislation like this every step of the way.

State Governments. Congress prohibits states from setting their own fuel economy standards, but many are considering "feebates." People who buy gas-guzzlers would pay extra fees...which would fund rebates for buyers of efficient cars. Maryland passed the first feebate law in 1992. Carmakers and the Bush administration plan to fight it in court.

- In the past, federal transportation funds were earmarked specifically for highways. But in 1991, Congress passed a law allowing states to spend some, or even most, of this money on energy-efficient mass transit—if they choose. Now it's up to governors and their state transportation agencies to make the best use of the money.

- Traditionally, the more energy we've used, the more money utilities have made. So, in general, they've never been enthusiastic about energy efficiency. But now, in some parts of the country, utility regulators—who are either appointed by governors or elected directly—are allowing utilities to keep a share of the money they save for their customers with conservation programs. The result: Instead of building power plants, these utilities are pursuing energy efficiency as their most important new "source" of energy.

Local Governments. If a utility is run by a city, the mayor or city council have the power to improve its conservation policy. For example: Santa Monica, California, Minneapolis, Minnesota, and Austin, Texas, have all provided free door-to-door energy audits for their residents. Auditors showed people how they could save more energy, and distributed simple energy-saving devices.

- Osage, Iowa's municipal utility has made energy conservation a priority for the last 20 years. Because of its emphasis on energy conservation, the community has cut its gas and electric consumption by 38% since 1974, saving residents $1.2 million a year in energy costs.

- City officials also have the power to promote transportation efficiency through carpool lanes, carpooling programs, expanding mass transit, and purchasing fuel-efficient vehicles for police and city fleets.

THE POLITICAL DEBATE: *Some arguments you might hear*

Argument: "The federal government is already doing a lot to promote energy conservation."

Answer: Not true. Except for a modest amount of increased funding for alternative-energy research and a few miscellaneous programs, federal programs have stressed maintaining current consumption levels of fossil fuels.

Argument: "Conservation isn't cost-effective—saving energy costs money."

Answer: Energy conservation measures have already saved America an estimated $160 billion a year since the OPEC oil embargo of 1973—and using today's technology, we could save an estimated $220 billion *a year*. True, we'll need to make some up-front investments—but the long-term payback will be as much as 20% to 50% every year.

Argument: "Requiring more fuel-efficient vehicles would drive U.S. automakers out of business."

Answer: Since the early '70s, U.S. carmakers have been saying that fuel-efficient cars aren't attractive to consumers, can't be made as safe as gas-guzzlers, and aren't profitable.

But at the same time that Detroit lobbyists and lawyers have been claiming that greater fuel efficiency is impossible, Honda's engineers have been building cars that already meet the challenge. The 1992 Honda Civic VX gets 55 mpg—50% more than the 1991 model.

Fuel-efficient cars, like all cars, *can* be engineered to be safer, with air-bags, stronger frames, and smarter design. But U.S. companies have never made that commitment to small cars.

And as for profitability—it's true that Detroit makes its biggest profits from selling its biggest cars, but only because it faces no competition from imports in the biggest size categories.

Argument: "We don't need to conserve energy to reduce our dependency on foreign oil. We should just drill in America."

Answer: Some people call this the "drain America first" approach. The U.S. now has less than 4% of the world's remaining oil reserves. (Saudi Arabia gets more oil out of 1,000 wells than we get out of 200,000 wells.) The only way to get more oil in the U.S. is to go to remote, environmentally sensitive, and expensive-to-drill-in areas like coastal waters and the Arctic Refuge.

But it's not worth the expense. Exploiting the Arctic Refuge, for example, would only produce about 6 months' worth of the U.S. demand for oil. When we've used that up, we'll still have to find more—but we will have destroyed one of the last truly wild places on the continent.

The only sensible solutions to our oil dependence are improving energy efficiency and developing renewable energy sources.

Argument: "We've got plenty of coal in America. Let's use that."

Answer: Burning coal contributes significantly to the greenhouse effect, air pollution, and acid rain. It's a massive health problem for miners, too. If we rely on it, we're poisoning ourselves...and hastening the destruction of the environment.

Argument: "Nuclear power is the solution—it's a clean, accessible source of energy."

Answer: Although nuclear power plants themselves don't emit any carbon dioxide or air pollution, making the fuel for nuclear reactors contributes substantially to the greenhouse effect. And nuclear waste is the worst hazardous waste problem on the planet. On top of that, the plants themselves are unsafe (consider the accidents at Three Mile Island and Chernobyl).

VOTE FOR THE EARTH: *Picking the right candidates*

√ Do candidates promote energy efficiency as our first, best energy resource—or do they treat it as an afterthought?

√ Do candidates support substantially higher fuel economy standards for cars and trucks?

√ Do they support expanded tax credits and research programs for renewable forms of energy?

√ Are state candidates in favor of making it profitable for utilities to invest in conservation?

√ Do local candidates support using federal transportation funds for mass transit, ridesharing, and bike lanes—instead of highways?

Voter Notes

• See if the candidate has a background in the energy business. It's not surprising, for example, that George Bush—who made a fortune in the oil business—hasn't supported energy conservation.

• Find out what kind of cars candidates drive. Are they gas-guzzlers?

VOTE TO SAVE ANCIENT FORESTS

Ninety percent of America's ancient forests have already been cut down.

By now, most people know about the fight between environmentalists and the logging industry over the fate of the last remaining old-growth forests in America. Unfortunately, it's been characterized as a choice between saving an obscure species of owl and saving jobs that support thousands of American families.

But it's really a choice between losing one of our last precious resources—*along with the jobs*…or finding a way to protect both the environment and local economies at the same time.

The Pacific Northwest was once covered with ancient forests. Now only a small number of these magnificent trees are left for our children…and their fate is in the hands of the people we elect to national office—our senators, our representatives, and our president.

FOREST FACTS

• "Ancient" forests contain trees that range from 200 to 1,000 years old. These include cedar, fir, pine, hemlock, spruce, and redwood.

• These trees—some of which are 350 feet tall—are the largest living things on Earth. There are no trees *anywhere in the world* like the ones in our ancient forests. They are part of a unique, irreplaceable ecosystem.

• Ancient forests in the U.S. provide habitats for more than 200 species of fish and wildlife that may otherwise face extinction.

• All but a tiny percentage of the remaining ancient forests are on land controlled by the U.S. Forest Service and the Bureau of Land Management—which means they're owned by the American people. Yet they're being sold to commercial lumber companies, often at a loss.

Going, Going...

• During the 1980s, nearly 200,000 acres of ancient forest were cleared every year. That's the size of almost 500 football fields a day. Experts estimate that at current cutting rates (which are half that of the 1980s), all of our ancient forests will be gone in less than 20 years.

• The remaining forests in the Pacific Northwest are protecting watersheds by holding soil in place and regulating water temperatures. When they're cut down, large amounts of soil erode into nearby rivers and lakes—killing fish and polluting water supplies.

• Scientists are just beginning to study and understand the importance of these trees. They now know, for example, that ancient forests may yield new medicines, like taxol (see Vote to Save Endangered Species, p. 60), which fights cancer.

• Old-growth forests play an important role in slowing global warming. They absorb three times as much carbon per acre as tropical forests, and store huge quantities of it. Because this carbon is released when the giant trees are cut down, logging actually contributes to global warming (see Vote to Stop Global Warming, p. 29).

WHAT OUR POLITICIANS ARE DOING: A *few examples*

The White House. In 1992, the Bush administration unveiled a plan to "protect" the northern spotted owl that ultimately guarantees its extinction—and the destruction of ancient forests.

Congress. More than 15 bills have been introduced to solve the ancient-forest crisis. Some may eventually be voted on. Environmentalists say the following points must be included in the final bill in order for it to be acceptable:

√ A reserve system to set the most vulnerable and ecologically significant areas of ancient forest off-limits to loggers.

√ New forestry practices, featuring tightly controlled, sustainable logging, for the areas surrounding the reserves.

√ Economic assistance for timber workers, similar to the GI bill—including education, training, and low-interest loans to help people who lose their jobs due to logging restrictions make a transition to another industry.

√ Restrictions on log exports. This might include a tariff and would at least mean exporters would have to pay income tax on logs sold abroad—something they don't do now.

• The Endangered Species Act is also up for reauthorization in 1992, and will probably be voted on in 1993.

THE POLITICAL DEBATE: *Some arguments you might hear*

Argument: "People's jobs are more important than owls."

Answer: "Jobs vs. owls" is a phony debate created by timber companies and anti-environmental politicians to confuse workers and voters. The real issue is whether or not we're going to save our last remaining ancient forests. The spotted owl is important mainly because it's considered an indicator of the health of the ancient forests—like a canary in a coal mine. Because the owl is dying out, we know that the entire forest ecosystem is endangered.

Argument: "The logging companies are protecting U.S. jobs."

Answer: Timber companies *say* they're fighting to protect jobs, but their practices show that, in reality, they're only concerned with profits.

• When we export logs, we're exporting jobs along with them. In 1988, 4.3 billion board feet of whole logs were exported from U.S. ports on the Pacific Coast—more than 25% of the total harvest from Oregon, Washington, and Alaska.

• Most of these logs were sent to Pacific Rim countries like Japan, Korea, and Taiwan to be milled there. So the lumber companies were actually creating jobs in Asia, not the U.S. If the logs had been milled at home, an estimated 17,000 jobs would have been created.

Argument: "Environmentalists are to blame for lay-offs in the logging industry."

Answer: No, automation is. A Forest Service study shows that even without stricter environmental controls, the number of timber-related jobs in the Douglas fir-region of Washington and Oregon will drop to around 50% of its 1970 level by the year 2000.

This has already begun to happen; although lumber production actually increased during the 1980s, employment went down.

• From 1977 to 1988, timber production in Oregon and Washington grew by 17%—but the number of workers fell by 19%.

• Since 1979, plywood production in Washington and Oregon increased by 10%. But because the mills were more efficient, more than 26,000 lumber jobs were lost.

• Twenty-five percent of all trees cut down during the 1980s were shipped overseas as whole logs.

• On top of everything else, most experts say that, at the present rate of logging, the ancient forests will be exhausted and loggers will be out of work in 10 to 20 years.

Argument: "Protecting these trees will destroy local economies."

Answer: Actually, the opposite is probably true. In fact, protecting the trees will help support the Pacific Northwest's long-term future. Consider this:

• Recreation and tourism, which depend on healthy forests and natural beauty, are among the fastest growing industries in the region, already worth more than an estimated $6 billion a year.

• Approximately 60,000 salmon-fishing jobs depend on clean water running through uncut forests. And the commercial fishing industry in the Pacific Northwest is worth $1 billion.

• Some domestic water supplies depend on healthy forests. Supplies for the Portland area, for example, originate in the watersheds of the Mt. Hood National Forest. The estimated worth of the water is about $24 million a year.

Currently, every industry in the Pacific Northwest—except timber, which represents less than 5% of the regional economy—is thriving. "In 1989," the Wilderness Society reports, "more than 160,000 new jobs were created in the region and labor experts predict that 1.6 million new jobs would be created by 2000." New jobs are created every year, and more companies are moving to the area. Undeniably, part of the attraction is its natural beauty, including the forests.

Argument: "Logging companies plant as many trees as they cut."

Answer: Trees, yes—forests, no. One hundred 6-inch saplings are not comparable to one hundred giant redwoods in any sense, but especially not in providing wildlife habitat. Tree farms are designed to be harvested in 50 to 80 years, long before they can develop the most important characteristics of an ancient forest. And they don't provide the habitat that forest creatures need to survive. In fact, experts say that these farms have more in common with a cornfield than with an old-growth ecosystem.

Plus, many of the trees planted by timber companies die before reaching maturity. A June 1992 Congressional study confirms that even when trees are replanted, they don't always grow back. "The federal government has exaggerated the success of tree-planting efforts," the report says.

VOTE FOR THE EARTH: *Picking the right candidates*

√ Do candidates regard environmental issues as jobs vs. nature—or do they understand that a good environmental policy is a good economic policy?

√ Will they support strong and effective ancient-forest protection legislation right away? We're losing valuable forests every day.

√ Do they support reauthorization of the Endangered Species Act and local endangered species legislation?

Voter Note

• Be wary of politicians who talk about "balance" or a "balanced resolution" to this crisis. We only have 10% of our forests left. It's too late for "balance."

HELPING

AT HOME

VOTE FOR RECYCLING

Every year, Americans produce about 180 million tons of solid waste.
That's about 4 pounds per person per day...and the figure is growing.

Everyone's heard about the "garbage crisis." We're producing too much trash, and all of the disposal methods we currently use can hurt both the environment and our health.

Even the "safest" landfills will eventually leak, threatening to contaminate the soil and water. And incinerators, promoted as the "high-tech" solution to the garbage problem, are spewing tons of poisonous chemicals into the air. What's more, everything we dump in a landfill or burn in an incinerator represents wasted resources we can't afford to squander.

That's where recycling comes in.

Americans have shown that they're willing to help solve the garbage crisis by recycling; every community with a recycling program has a high participation rate. The problem is, our federal, state, and local governments haven't all made that same commitment.

The solution: We need to elect candidates who are willing to make recycling a priority.

DID YOU KNOW

• 70% of America's 20,000 landfills closed between 1978 and 1988. By 1993, another 2,000 are expected to close.

• Many state and local governments are turning to incinerators as a "quick fix" solution to the garbage crisis. But incinerating only reduces trash by 70% and leaves millions of tons of toxic ash behind.

• According to the EPA, only about 13% of our municipal solid waste is recycled—even though recycling saves money and resources *and* reduces pollution. For example:

√ Making cans from recycled aluminum cuts related air pollution by 95% and saves 95% of the energy used to make new cans.

√ More than a ton of resources is saved for every ton of glass that's recycled.

√ Every year, enough energy is saved by recycling steel to supply Los Angeles with almost a decade's worth of electricity.

WHAT OUR POLITICIANS ARE DOING: *A few examples*

The White House. In 1990, the EPA proposed an ambitious recycling initiative. But the Council on Competitiveness (chaired by Vice President Dan Quayle) rejected it, on the grounds that it competed with the incinerator industry.

• Now the EPA has to figure out what to do with millions of tons of incinerator ash—which contains lead, mercury, and other toxic heavy metals. It's considering using the ash for road surfacing, which is dangerous because it can pollute soil and water.

Congress is currently revising the Resource Conservation and Recovery Act (RCRA), which is supposed to offer a comprehensive plan for solving the U.S. garbage crisis.

• But because of waste-industry lobbying, many provisions that would promote recycling over other options (such as incineration) have been cut. For example: There are few requirements for companies to use recycled materials, no national bottle bill, and no limit on incineration. If our representatives were more committed to recycling, those provisions would still be included.

State Governments. Electing pro-recycling candidates pays. More than half the states have already passed laws to stimulate recycling—from packaging-reduction laws to bottle bills. California's new recycling law requires newspapers to contain 50% recycled content by the year 2000. A few states have even decided not to build new incinerators, to give their recycling industries a chance to grow.

Local Governments. By October 1990, an estimated 75 cities were recycling more than 25% of their waste. Some cities are recycling even more; for example: Seattle, Washington (36%), Longmeadow, Massachusetts (49%), and Berlin, New Jersey (57%).

THE POLITICAL DEBATE: *Some arguments you might hear*

Argument: "Recycling doesn't work. All those materials we recycle just get thrown out, anyway."

Answer: Unfortunately, it's true that materials are sometimes dumped in landfills instead of being recycled.

But that's not because recycling doesn't work—it's because, under current government policy, it's easier and cheaper for manufacturers to use virgin materials than recycled materials. The problem: Rather than providing businesses with incentives to use recycled materials, our tax system actually subsidizes the use of virgin materials.

For example: If you're cutting down timber for paper, you get tax breaks for the depletion of the forest. But you get no tax breaks for building a plant that processes recycled fiber or newsprint. That means taxpayers wind up *paying* businesses to use trees instead of recycled paper.

Taxpayers also pay for garbage collection and disposal. So a company can use extra packaging that produces extra trash, yet never have to take responsibility for the expense it's created.

That's not right. If you look at collection and disposal as part of the cost of the packaging—which it is—then it makes sense for manufacturers, instead of taxpayers, to pay for the hidden costs of using virgin glass, plastic, paper, and other recyclable materials. If we give tax breaks to companies that *reuse* material instead of those that create new garbage, we'll be using the tax structure to encourage recycling and creating new markets for recycled materials.

Argument: "Incinerators are a terrific solution. By burning garbage we not only get rid of it, we turn it into energy at the same time."

Answer: Generating energy by burning garbage in incinerators (called "waste-to-energy" by the waste industry, and "wasted energy" by environmentalists) sounds like a good idea, but it isn't.

First of all, a lot of recyclable materials burned in incinerators are not "burnable." That means it takes more energy to burn them than we get back *from* burning them.

Second, even burnable things like paper or plastic are a net energy loss compared to recycling, because you save more energy by recycling them *just once* than you create by burning them.

On top of all that, incineration doesn't eliminate garbage—it merely reduces the volume, leaving behind toxic ash. "An incinerator takes 1,000 pounds of relatively harmless garbage," says Daniel Weiss of the Sierra Club, "and turns it into 300 to 500 pounds of poison ash." That's not a very good trade-off.

There's more: Garbage incineration is the most expensive way ever developed to get rid of trash. The city of Detroit, for example, built a huge garbage incinerator. Now the city has to divert money from its police and fire departments to pay the debt incurred from building and operating it. On the other hand, the city of Seattle put the money it was going use for a garbage incinerator into developing the nation's best big-city recycling program. The result: Seattle is recycling 36% of its garbage without the cost overruns, toxic exposure, and other problems that come with incinerators.

So who wants us to use incinerators? Waste-management companies. They make profits from building and operating these facilities.

Argument: "These are scare tactics—the waste from incinerators is harmless, and recycling won't get rid of all our garbage."

Answer: It's true that toxic ash doesn't always flunk the EPA test that determines whether a particular waste is officially classified as "hazardous"...but it can still be contaminated with lead, mercury, dioxins, and other dangerous substances. In this case, "hazardous" is just a legal term, not an indication of whether the ash is harmful.

As for recycling replacing incineration: There are existing or planned garbage incinerators in nearly every state. So we have enough already built. Let's concentrate on recycling and use the unrecyclable garbage to fuel our existing incinerators.

Argument: "People won't recycle more just because of a regulation."

Answer: More than 70% of the glass and virtually all of the plastic recycled in the U.S. is collected in only nine states. The reason: These states have passed bottle bills. This demonstrates that laws really can boost recycling. If the federal government passed a national bottle bill, our national recycling rate could easily double.

VOTE FOR THE EARTH: *Picking the right candidates*

√ Do they support a national or state bottle bill?

√ Are they willing to require manufacturers to use recycled materials in their new products?

√ Do local candidates support curbside recycling programs?

√ Do they have a recycling program in their own offices?

√ Is their campaign literature printed on recycled paper?

VOTE FOR SAFE FOOD

U.S. farms use an estimated 845 million pounds of pesticides every year.

Americans are increasingly health conscious...but today, even the most nutritious foods may not be healthy. Since 1945, pesticide use in this country has increased tenfold, and residues from these highly toxic chemicals now permeate our food supply.

For example: Apples—the all-American symbol of good health —can legally be sprayed with as many as 26 cancer-causing pesticides. Tomatoes can be sprayed with 23. And the fish in our lakes and rivers are exposed to hundreds of these dangerous chemicals from polluted runoff.

We need representatives who will work to keep pesticides out of our environment...and out of our food.

FOOD FACTS

• According to the EPA, about 1 in 5 pesticides currently used on food in the U.S. may cause cancer in human beings. Yet, the agency has banned only 11 of the 69 most commonly used carcinogenic pesticides.

• In 1988, an FDA survey of childhood exposure to pesticides found "detectable" levels of pesticides in 1/3 to 1/2 of all meat, fish, fruit, and vegetables.

• Pesticides banned or unregistered in the U.S. can still be exported and used on crops grown in other countries. This creates a "circle of poison": Banned pesticides are used on fruit that is imported into the U.S....and, although the pesticides are illegal, we consume them. Foreign-grown fruit and vegetables now account for 25% of all produce Americans eat...and an estimated 5% of imported produce is contaminated with banned pesticides like DDT.

• Seafood is the number-one source of toxic and cancer-causing chemicals in the U.S. food supply today. In a two-year study, 70%

of the U.S. seafood tested was contaminated with pesticides.

WHAT OUR POLITICIANS ARE DOING: *A few examples*

The White House. In 1990, the Bush administration, backed by the pesticide lobby, defeated an attempt by Congress to ban the export of pesticides that are illegal in the U.S.

Congress. The first law regulating pesticides—the Federal Insecticide, Fungicide and Rodenticide Act (FIFRA)—was passed in 1947. The law was amended in 1972. In 1992, FIFRA comes up for reauthorization, and legislators have a chance to implement even tighter controls. Other legislation specifically controlling pesticides in foods has also been introduced.

• Many states and hundreds of local communities (see below) have passed laws regulating the use of pesticides. But this right is in jeopardy. If the House's Pesticide Safety Improvement Act of 1992 passes, it will revoke the right of state and local governments to regulate pesticides.

The EPA (Executive Branch). Many pesticides now in use were originally registered in the '40s, '50s, and '60s, when safety standards were lower. Under a 1988 law, manufacturers have until 1997 to reregister these "old" pesticides with the EPA, to show that they meet current standards. But the EPA is not ensuring compliance. As of April 1992, the EPA had registered just 14 of more than 600 pesticides requiring action. It now says reregistration will continue into the next century.

State and Local Governments. The 1990 Farm Bill required the federal government to set up a national board to certify organic produce by 1993, but so far little action has been taken. However, 26 states have enacted their own "organic" labeling or certification laws.

• The state of Iowa protected its groundwater supplies by passing a law that greatly restricted the use of atrazine—a common agricultural pesticide.

• And the city council of Lebanon, Maine passed a law banning the use of pesticides throughout the city—on roadsides, golf courses, schoolgrounds, city parks, etc.

THE POLITICAL DEBATE: *Some arguments you might hear*

Argument: "Without using pesticides, we can't grow enough food."

Answer: That's what a lot of people have been led to believe, but it's not true. The fact is, pesticides are becoming *less* effective. While pesticide use has grown tenfold in the last 45 years, crop losses due to pests have almost doubled—from 7% to 13%. The main reason: Over time, an increasing number of insects, plant diseases, and weeds have become resistant to the pesticides we use. Insect resistance has increased from 137 species in the early 1960s to more than 500 currently.

On the other hand, an increasing number of studies show that using safer "low-tech" and organic alternatives—like crop rotation and the development of pest-resistant plant strains—does *not* reduce crop yields. Some experts even estimate that the U.S. could decrease its pesticide use by 50% without affecting crop production. (Sweden has successfully cut its pesticide use in half since 1985...and has set a similar new goal for 1995.)

Alternative pest controls can even save U.S. farmers money. In a Pennsylvania study, organic techniques earned farmers an overall profit of $345 per acre, compared to a net loss of $61 per acre when techniques relying on pesticides were used.

Argument: "All that talk about pesticides causing cancer is just a scare tactic. The risks are exaggerated by test results; humans actually get a much lower dose of pesticides than lab animals."

Answer: Scientists agree that while laboratory tests aren't perfect, they're still the best indication of whether the toxic chemicals in pesticides are harmful to people. What's more, their findings are often backed up by human studies, which show increased cancer rates among farm and chemical workers regularly exposed to pesticides.

If test results are inaccurate, they're more likely to *underestimate* the hazards: government risk assessments are based on the assumption that people are exposed to only one toxic chemical at a time. The truth is, we are routinely exposed to a "soup" of toxic chemicals. For example, a study by the U.S. Public Interest Research Group found that the food in a normal day's menu could contain residues from as many as 60 carcinogenic pesticides.

Argument: "U.S. consumers are used to 'perfect'-looking fruits and vegetables. They won't buy produce that doesn't look good."

Answer: The idea that fruits and vegetables must look picture-perfect to be good for you has been marketed to the American consumer only since World War II. Yet today, government size and shape requirements for produce account for 40% to 60% of all pesticide use in the U.S. The truth is, a few surface bumps and bruises have little relation to how produce tastes or its nutritional value. And studies show that both farmers and consumers favor reducing pesticide use, even if it means less-than-perfect-looking produce.

Argument: "If a pesticide is registered with the EPA, it's safe."

Answer: The EPA's pesticide registration process is supposed to weigh the economic benefits of pesticides against the health risks. But when a company spends money on developing a pesticide, the EPA often allows it to put the new pesticide on the market before it's even tested. And then the pesticide is removed only if there's proof of a problem.

But, even *that* may not result in removal from the market. One current example: The pesticide 2,4-D is a proven carcinogen. This has been shown in the results of at least four studies, including two by the National Cancer Institute. But apparently that's not enough. The chemical is still on the market.

VOTE FOR THE EARTH: *Picking the right candidates*

√ Do candidates acknowledge the potential dangers and health risks associated with pesticide use and exposure?

√ Are they in favor of setting stricter pesticide standards in their communities or states?

√ Do they support subsidies for organic farming (national and state candidates) or farmer's markets (local)?

Voter Note

• If candidates are supported by pesticide industry-oriented groups like the Coalition for Sensible Pesticide Policy, the American Council on Science and Health, and the National Agricultural and Chemical Association, or any chemical manufacturers, they are not likely to protect your food.

VOTE FOR CLEAN AIR

The EPA estimates that more than 2.7 billion pounds of toxic chemicals alone are released by the U.S. into the air every year.

The brown haze hanging over our city skylines…the exhaust from your car…the emissions spewed from power plants and factories: this pollution isn't just ugly; it's deadly.

Nothing is more precious than the air we breathe…But we still poison it by releasing millions of pounds of dangerous chemicals into it every day. As a result, according to the EPA, more than 100 million Americans now live in places where U.S. clean-air standards are violated some time during the year.

No American should have to live where the air is unhealthy. We need to elect people who understand this…and are willing to fight to clean up our air—and keep it clean.

DID YOU KNOW

• Air pollution is a broad term that refers to:

√ Smog, which is created when gases produced by burning fossil fuels (gasoline, coal, natural gas) are "baked" by the sun.

√ Toxic chemicals (including carcinogens), which are released into the air by refineries, chemical plants, and other industries.

√ Nitrogen oxides and sulfur oxides (gases released into the air by coal-burning electric power plants and motor vehicles), which mix with rain or snow in clouds to form acid rain.

√ Carbon monoxide, a deadly gas released into the air when fuel doesn't burn completely. (Cars and trucks are the main source.)

• According to the American Lung Association, smog, sulfur dioxide, and airborne toxics cause more than 120,000 deaths a year and cost the U.S. as much as *$100 billion* a year in health care and lost productivity.

• The U.S., which has only 5% of the Earth's population, produces 70% of its carbon monoxide gas, 23% of the carbon dioxide, 45% of the nitrogen oxides, and 34% of the hydrocarbons (a cause of smog).

- There are approximately 140 million cars and trucks on the road in the United States. The exhaust they produce is responsible for about 50% of our air pollution.

- Acid rain in the U.S. has killed marine life in lakes and streams, damaged forests, contaminated drinking water, and even eroded buildings. One example: 36% of the lakes in the Adirondack Mountains in New York State are dying from acid rain.

- Air pollution causes respiratory diseases and can weaken even healthy hearts. For a person who has heart disease, it can be fatal.

WHAT OUR POLITICIANS ARE DOING: *A few examples*

The White House. The Council on Competitiveness—appointed by the president and chaired by Dan Quayle—has intentionally undermined the 1990 Clean Air Act (see The Political Debate).

Congress. The Clean Air Act, originally passed in 1970, was the first U.S. law designed to regulate air pollution. Congress updated the law in 1990. But while the law has been toughened, it still includes plenty of loopholes. For example: It doesn't regulate toxic emissions from utilities.

The EPA (Executive Branch). Under the original Clean Air Act, the EPA was supposed to regulate toxic airborne chemicals. But it has set standards for *only* 7 of the hundreds of potentially toxic chemicals—and hasn't set *any* standards since 1990.

State Governments. Some states have passed air pollution laws that are tougher than federal laws. For example: California has its own auto emission standards that require carmakers to produce more efficient cars. And Colorado has proposed tougher enforcement of air pollution laws to include jail terms for industrial polluters.

Local Governments. Some cities and counties fight air pollution with "automobile trip reduction" laws, which require larger companies to reduce the number of commuters traveling alone by promoting carpooling, mass transit use, and bicycling. Not only do these measures reduce pollution, but companies benefit as well (e.g., from reduced parking expenses). These ordinances have been successful in Montgomery County, Maryland, and Bellevue County, Washington.

THE POLITICAL DEBATE: *Some arguments you might hear*

Argument: "The new federal Clean Air Act will take care of the problem."

Answer: How can it, when it isn't being enforced? After calling the Clean Air Act the "centerpiece" of his environmental program, President Bush has done his best to keep it from having any effect. He called for a "moratorium" on regulations last February. And he gave Dan Quayle's Council on Competitiveness—whose purpose is to "stimulate the economy by promoting *de*regulation"—veto power over EPA guidelines.

Now the council can undermine the law simply by withholding approval of guidelines until the EPA rewrites them in industry's favor. For example: The council rejected a ban on the burning of toxic lead batteries and added a loophole to the section of the law that requires polluters to get permits before polluting. The permits are supposed to be applied for in public, with a hearing and an opportunity for "judicial challenge." Now companies can apply to increase their air pollution by up to 490,000 pounds a year, *without public notification*. And they don't even have to wait for approval. While the application is pending, a company has the right to go ahead with its plans.

As a result of the council's interference, the EPA has missed at least two deadlines for setting guidelines for airborne toxics. Because of the delay, one billion extra pounds of toxics have been released into the air. Not only is this immoral, it's illegal. And environmental groups are suing.

Argument: "Air pollution controls are too expensive."

Answer: Industries have been complaining about the cost of cutting their pollution for decades. In 1970, for example, Lee Iacocca warned that the original Clean Air Act could "prevent the continued production of automobiles." Now industry is making similarly inflated claims.

Of course, it will be expensive to clean up our air. The EPA and the President's Council of Economic Advisors estimate that the new Clean Air Act will cost businesses about $25 billion more per year when it's fully implemented. But right now the price of dirty air is being paid by the rest of us—in health problems and in damage to our crops, lakes, streams, and forests. Crop damage from acid rain alone is estimated to cost $2 billion a year. And the American Lung Association reports that we lose $100 billion a year in health costs and productivity.

The trade-off already seems more than worth it. But there are additional benefits: The $25 billion spent each year wouldn't be a loss to our economy—it would be an investment in the American clean up industry...and would produce thousands of jobs. In addition, American companies that cut their pollution often find it *saves* them money. One example: In 1987, Fisher Controls International, an industrial equipment manufacturer in Iowa, was releasing 100,000 pounds of airborne toxics, was producing 870,000 pounds of hazardous waste, and was shipping more than 1,400 tons of garbage to the local landfill. By 1991, the company had reduced its air emissions by 80%, slashed hazardous waste by 90%, and cut garbage going to the landfill by half. This year its cleanup program will save the com-pany about $1.4 million.

Argument: "If we're going to have factories, we're always going to have some kind of air pollution."

Answer: No one's talking about getting rid of air pollution—just getting it down to tolerable levels. Consider this: The acid rain regulations under the new Clean Air Act are only expected to cut acid rain by about 40%. And some of the other controls won't go into effect until 2023. That's certainly not going overboard. The real question is, is it *enough?*

VOTE FOR THE EARTH: *Picking the right candidates*

√ Are they in favor of reducing air pollution by increasing automobile fuel efficiency?

√ Look for state candidates who will fight to have their state adopt California's more stringent low-emission auto standards.

√ Local politicians should be willing to pass laws that promote bicycle use, carpooling, and mass transit, as well as housing and land-use projects that keep jobs close to residential neighborhoods.

Voter Note

• Candidates who are strongly associated with automobile, coal, or chemical companies may not support clean-air legislation. These industries have worked against clean-air laws.

VOTE FOR CLEAN DRINKING WATER

The EPA reports that 50% of public water supplies tested nationwide show some level of toxic contamination.

Most of us assume the water that comes from our faucets is relatively clean and safe to drink.

Unfortunately, we may be wrong.

According to the EPA, toxic pollution of drinking water is one of the greatest environmental hazards in the United States today. Every year, the health of millions of Americans is put at risk by contaminated drinking water. Children are especially at risk because they weigh less and have less developed, more sensitive nervous systems.

Who has the power to protect you and your family? Your elected officials. Some pollutants—radioactive contaminants; toxic chemicals dumped in lakes and rivers; pesticide runoff; lead in plumbing pipes; and so on—are already regulated by government. But the regulations are not strong enough and are rarely enforced.

The right person in the mayor's or governor's office...the legislature...or even the White House could keep polluters from contaminating your drinking water.

DID YOU KNOW

• More than half of Americans get their drinking water from natural wells and aquifers ("groundwater"). The rest get it from rivers, lakes, and reservoirs.

• Between 1986 and 1988, the Environmental Protection Agency and the Centers for Disease Control recorded more than 25,000 cases of illness from contaminated drinking water.

• Contaminated drinking water has been linked to nerve damage, kidney disease, cancer, and lower intelligence in children. In a few cases, people have even died from it.

Water Pollution

• Groundwater is becoming increasingly polluted. According to one report, the EPA has found at least 74 pesticides in the groundwater of more than 38 states. And once groundwater is contaminated, it's very expensive—and sometimes impossible—to clean it up.

• Underground tanks in the U.S. are currently leaking gasoline and other hazardous chemicals into the soil...where they can seep into groundwater supplies. A single gallon of gasoline can pollute 750,000 gallons of water.

• In 1989, industries dumped an estimated 189 million pounds of toxic chemicals into U.S. waterways.

• According to recent studies, as many as two-thirds of the nation's rivers are contaminated by pesticide runoff from farms.

• Although Congress has banned lead water pipes in new construction, an estimated 130 million Americans are still being exposed to hazardous levels of lead in their drinking water. The reason: EPA regulations allow long intervals for replacing lead pipes (as long as 24 years).

WHAT OUR POLITICIANS ARE DOING: *A few examples*

Congress. The Clean Water Act, passed in 1972, was supposed to eliminate all U.S. water pollution by 1985. It *has* improved the quality of drinking water, but it needs to do more: the most recent EPA figures show that river pollution in 35 states and lake pollution in 16 states still violates clean-water standards.

• In 1974, Congress passed the Safe Drinking Water Act. Although it has reduced microbiological contamination, contamination of drinking water from pesticides, toxics, and lead pipes has continued.

• Both the Clean Water and Safe Drinking Water acts are up for reauthorization in 1992. We need Congress to strengthen both.

The EPA (Executive Branch). More than 100,000 violations of the Safe Drinking Water Act are reported to the EPA each year. But the law is not being enforced: in 1990, less than 15 federal cases resulted in fines, and only 18 violations were referred to the Justice Department for possible prosecution.

• Even when fines *are* levied, they are often so low that they have little impact.

State and Local Governments. The Safe Drinking Water Act requires states to tell the EPA whether their water supplies meet water-safety testing requirements; but some states don't, and, according to a government study, some water suppliers even falsify their reports. However, some states are passing their own clean-water initiatives: a 1990 New Jersey law requires polluting companies to pay fines at least as high as the profits they've made from dumping the hazardous wastes.

• Some federal water-safety testing standards only apply to water systems serving 10,000 people or more. This means the health of millions of Americans who live in rural areas and small towns is threatened. It's up to state and local governments to protect these people.

THE POLITICAL DEBATE: *Some arguments you might hear*

Argument: "Our drinking water is safe—it meets EPA standards."
Answer: In the first place, drinking water in more than 15,000 water systems (for example, Long Island, New York) doesn't meet EPA regulations...and the people in those communities may not know it.

In the second place, the EPA doesn't *have* any standards yet for many of the most toxic pollutants. The Safe Drinking Water Act required the EPA to set national safety standards for 83 additional hazardous water-contaminating chemicals by 1989. As of 1991, standards for only 50 of these chemicals had been set.

And finally, many of the current EPA guidelines aren't strong enough. For example, the EPA standard for arsenic in drinking water—50ppb—puts people at a significantly increased risk of cancer.

Argument: "We have to weigh the costs of having safe drinking water against the benefits. Some risks may be acceptable."
Answer: Every person deserves clean drinking water. Risk assessment is a poor excuse for allowing continued pollution of our water supplies—especially for those who may develop cancer or other life-threatening diseases as a result. If someone in your family became ill from drinking contaminated water, would you consider the risk acceptable?

Argument: "Cleaning up our water will cost American businesses jobs, efficiency, and competitiveness. We can't afford it."

Answer: Contaminated water now *costs* the U.S. economy more than $10 billion a year in cleanup costs. *That's* what we can't afford.

On the other hand, preventing pollution will save money and create jobs. For example, upgrading sewers, treatment plants, and other water-related infrastructures would put people to work in construction, engineering, and transportation.

Furthermore, water contamination and waste water treatment are huge international problems. Technology and skills developed by the U.S. to clean up our water can be exported—creating thousands of jobs and making millions of dollars for U.S. companies.

VOTE FOR THE EARTH: *Picking the right candidates*

√ Do they support strengthening the Clean Water Act and Safe Drinking Water Act during reauthorization?

√ Are they willing to make polluters pay fines large enough to discourage them from polluting?

√ Do they support local clean-water legislation?

Voter Note

• Watch out for candidates who try to shift the responsibility for reducing water contamination from industries to municipal water systems...or even to homeowners who haven't replaced their lead pipes. This is a good sign that the candidate might fight clean-water laws in order to protect polluting companies.

VOTE FOR "GREEN" PRODUCTS

Polls show that 4 out of 5 people will choose a product that is environmentally safe over one that isn't.

After Earth Day in 1990, millions of Americans began buying "green" products to help protect the environment.

Now we spend billions of dollars a year on items labeled "recycled," "recyclable," "biodegradable," "nontoxic," "kind to the Earth," "Earth friendly," "ozone friendly," or just plain "green."

But believe it or not, we may be wasting our time shopping for eco-products…because legally, these terms mean nothing at all. Anyone can claim a product is "Earth friendly"…no matter what's in it.

Using our power as consumers is one of the most important ways we, as individuals, can help save our planet. But to be effective, we need legislators who will protect us from deceptive advertising.

DID YOU KNOW

• "Green" marketing is a growing part of U.S. business. In 1991, 12% of all *new* products that came out claimed to be good for the planet.

• Marketing experts estimate that by 1996 products claiming to be "green" will make up 20% of the entire consumer marketplace.

• However, many products don't live up to their environmental claims. For example:

√ Aerosols are sometimes labeled "ozone friendly." Apparently this means no ozone-depleting CFCs are used. But some "ozone-friendly" aerosols still contain a solvent called methyl chloroform…which also damages the ozone layer.

√ While most products that claim to be "recyclable" can *theoretically* be recycled, many would have to be transported more than 1,000 miles to reach a recycling center that could accept them.

√ Calling a product "recycled" implies that it's made from materials that have already been used by consumers. However, current laws allow manufacturers to call factory scraps—which would have been used again anyway—"recycled."

√ According to the Local Government Commission, labeling a product "nontoxic" can be misleading: "Manufacturers may place the word 'nontoxic' on their label simply by meeting the federal regulatory definition," it says. "This can mean, for example, that if less than 50% of lab animals die within two weeks when being exposed to the product through ingestion or inhalation, the product can be called 'nontoxic.'"

WHAT OUR POLITICIANS ARE DOING: *A few examples*
Congress. The proposed Environmental Marketing Claims Act, currently being debated in the House and Senate, would give the EPA the authority to set standards for four terms—"Recycled," "Recyclable," "Reusable," and "Biodegradable"—and prosecute companies that misuse them.

• According to the act's sponsors, many businesses object to it. One reason: Federal standards would only be a minimum, and individual states could pass stiffer regulations.

State and Local Governments. As of February 1992, eight states had laws regulating some environmental product claims—and many other states will introduce laws this year. However, some of these laws (e.g., Indiana's) set standards well below those proposed in the federal Environmental Marketing Claims Act.

• Some state governments have already begun to prosecute environmental claims cases under "false-advertising" laws. For example: In 1990, the attorney general of Minnesota led a suit against Mobil Oil for claiming that its Hefty trash bags were "biodegradable." Mobil settled the suit and took the "biodegradable" label off its bags.

• Local governments can act as well. For example: The Consumer Affairs Department of New York City filed a suit against Proctor & Gamble for running an ad that showed a pile of compost under the headline: "30 days ago this was a diaper." This was grossly misleading, and Proctor & Gamble ultimately pulled the ad—not only in New York, but nationwide.

THE POLITICAL DEBATE: *Some arguments you might hear*

Argument: "If the government regulated claims on 'green' products, it would be interfering with the free market."

Answer: Actually, the opposite is true. Green consumerism supports the free-market system by providing people with the information they need to make intelligent choices.

What's more, the government wouldn't be interfering in anything—its *only* role would be to define terms and prevent false and misleading advertising—a standard government practice for most of this century.

Argument: "Regulating products' environmental claims with a new law will cost the taxpayers too much money."

Answer: The government wouldn't be *regulating* environmental claims—it would be defining terms for manufacturers and taking enforcement action against people who break the law. And in the case of enforcement, the government would actually be making money rather than spending it, because fines would be collected.

Another important consideration is the cost to consumers of *not* regulating environmental claims. As the environment deteriorates, we pay increased health costs, emergency cleanup costs, etc. So taking steps now to protect the environment and give legitimate green consumerism a fair chance in the marketplace will benefit the economy.

Argument: "States shouldn't be allowed to set tougher standards than the federal government. It makes complying with the different rules too complicated for companies."

Answer: In other areas of commerce, states often pass their own regulations, and manufacturers are able to comply.

The truth is, industries that object to states having "preemptive" rights are just worried about their own profits. Federal standards are minimal, anyway. If industries really only want one set of standards, they should be willing to accept much tougher definitions. But beyond that, when states pass tough environmental laws, they serve as a model of what can be done. Many of our most important federal laws began in cities or states. This is the real reason industries object to allowing states the right to autonomy.

VOTE FOR THE EARTH: *Picking the right candidates*

√ Have they demonstrated a commitment to consumer protection in general—either by supporting pro-consumer legislation or speaking out for consumer rights?

√ Do they support tough penalties against people who mislead the public with false advertising claims?

√ Do they support the Environmental Marketing Claims Act (national candidates) or the state or local equivalent?

Voter Notes

• Using the terms "First Amendment right" or "free speech rights" in the context of green consumerism generally means a candidate opposes truth-in-packaging regarding products' environmental impact.

• Check to see if candidates are green consumers in their own lives and campaigns. For example: Is their campaign literature printed on recycled paper?

CANDIDATE

RECORDS

THE 1991 NATIONAL ENVIRONMENTAL SCORECARD

Every year, the League of Conservation Voters (LCV) publishes a *National Environmental Scorecard* showing how members of the U.S. Congress stand on important environmental bills. For 1991, LCV evaluated Congresspeople on 15 Senate bills and 13 House bills. Each member was then given a "score" based on his or her pro-Earth position.

In the following section, we've included:
• The overall 1991 LCV scores for each member of Congress (listed under "LCV '91 SCORE").
• The specific voting records of each member of the House and Senate on five sample environmental bills picked by the LCV (listed under "SAMPLE BILLS").

Here are descriptions of the bills:

SENATE BILLS

1. *National Energy Policy*. The Gulf War spurred a legislative push for a long-needed national energy policy. Unfortunately, a massive bill introduced by Senators Johnston (D-LA) and Wallop (R-WY) stressed increased development of fossil fuel and nuclear energy and offered only token efforts to increase energy efficiency.

In one of the most significant environmental victories of 1991, a filibuster prevented that bill from reaching the floor for a vote. A motion to invoke cloture (to end the filibuster so the bill could be voted on) was rejected 50-44. "No" votes are considered pro-environment votes, as are abstentions (supporters of the bill needed 60 "yes" votes to bring the legislation to the floor).

2. *Arctic Wilderness*. Senators Roth (R-DE) and Baucus (D-MT) introduced Senate Bill 39 to designate the 1.5-million-acre Arctic coastal plain of Alaska's Arctic National Wildlife Refuge as wilderness. Conservationists say this will protect the habitat of the

Porcupine Caribou herd and other species and preserve the wilderness value of the area. Representatives of the oil and gas industries argue that the area should be made available for exploration and development. Co-sponsors of this bill get pro-environment ratings.

3. Wetlands Protection. Senate Bill 1463 (introduced by Senator Breaux (D-LA) would gut Section 404 of the Clean Water Act and endanger millions of acres of wetlands across the country. Co-sponsors of this bill receive anti-environment ratings.

4. Population Stabilization. In 1986, the United States cut off all funding for the United Nations Population Fund (UNPFA), charging that it funded coercive programs (e.g., requiring mandatory abortions for women after they have one child) in China. UNPFA officials denied this charge.

An amendment introduced by Senator Simon (D-IL) would have authorized $20 million for UNPFA for contraceptive supplies with the stipulation that if any of the funds go to China, the entire $20 million would be returned to the U.S. The vote was on whether to invoke cloture (limit debate) so the amendment could come up for a vote. A "yes" vote is a pro-Earth vote.

Note: The motion was passed 63-33 (60 votes are required for cloture). The legislation then passed with the amendment intact, but the House-Senate conference modified it to require that the U.S. ambassador to the United Nations approve it.

5. Federal Lands. The federal government charges less than market rates for grazing livestock on nearly 270 million acres of public lands managed by the Bureau of Land Management and the U.S. Forest Service. For example, in 1991, those agencies charged $1.97 per animal unit month (AUM); the market value on equivalent private land is $9.00 per AUM. This means taxpayers are underwriting cattle ranching.

The Jeffords-Metzenbaum Amendment to the Fiscal Year 1992 Interior Appropriations bill would require the federal government to charge market value for grazing livestock.

Senator Domenici (R-NM) moved to kill the amendment. It was tabled with a vote of 60-38. A "no" vote is a pro-environment vote.

HOUSE BILLS

1. *Toxic Chemicals*. The Community Right to Know More Act of 1991 (House Resolution 2880, sponsored by Rep. Sikorski, D-MN 6) would require industries to report on toxic chemicals used and produced in addition to those released into the air and water. Co-sponsors of this bill get pro-environment ratings.

2. *Arctic Wilderness*. This is the House version of the Senate's Arctic Wilderness bill (see #2, p. 94). Rep. Robert Mrazek (D-NY 3) re-introduced House Joint Resolution 239 to designate the Arctic coastal plain as wilderness. Co-sponsors of the resolution get pro-environment ratings.

3. *Wetlands Protection*. House Resolution 1330—proposed by Rep. Hayes (D-LA 7)—would gut Section 404 of the Clean Water Act (see #3, p. 95). Co-sponsors get anti-environment ratings.

4. *Population Stabilization*. Rep. Smith (R-NJ 4) introduced this proposal to strip the $20 million originally provided for UNPFA (see #4, p. 95). Rep. Kostmayer (D-PA 8) offered a substitute amendment to ensure restored funding for UNPFA. The Kostmayer Amendment was adopted 234-188. A "yes" vote is a pro-environment vote.

5. *Grazing Fees*. An amendment similar to the Federal Lands Senate bill (see #5, p. 95) was introduced by Rep. Synar (D-OK 2) to raise grazing fees to fair market value. It was adopted 232-192, but was later dropped. A "yes" vote is a pro-Earth vote.

SCORECARD KEY

LCV '91 =		Voting record on 1991 environmental bills
SCORE		(15 in the Senate, 13 in the House) chosen by the League of Conservation Voters. Each pro-environment vote is approximately 7% of the total; 100% is possible.
+	=	Vote for the environment
—	=	Vote against the environment
Dst	=	District
*	=	Senate member up for reelection (All house members are up for reelection.)
I	=	Ineligible to vote

ALABAMA

SENATE

SENATE	LCV '91 SCORE	Energy	Arctic	Wtlnds	Pop	Lands
Heflin (D)	7%	—	—	+	—	—
Shelby* (D)	13%	—	—	—	+	—

HOUSE	Dst	LCV '91 SCORE	Toxic	Arctic	Wtlnds	Pop	Grazing
Bevill (D)	4	23%	—	—	+	—	—
Browder (D)	3	38%	—	—	+	—	—
Callahan (R)	1	0%	—	—	—	—	—
Cramer (D)	5	46%	—	—	+	+	—
Dickinson (R)	2	8%	—	—	—	+	—
Erdreich (D)	6	62%	—	—	+	+	+
Harris (D)	7	31%	—	—	—	—	—

ALASKA

SENATE

SENATE	LCV '91 SCORE	Energy	Arctic	Wtlnds	Pop	Lands
Murkowski* (R)	20%	—	—	—	+	—
Stevens (R)	20%	—	—	—	+	—

HOUSE	Dst	LCV '91 SCORE	Toxic	Arctic	Wtlnds	Pop	Grazing
Young (R)	1	0%	—	—	—	—	—

ARIZONA

SENATE

SENATE	LCV '91 SCORE	Energy	Arctic	Wtlnds	Pop	Lands
DeConcini (D)	33%	—	—	+	+	—
McCain* (R)	33%	—	—	+	—	—

HOUSE	Dst	LCV '91 SCORE	Toxic	Arctic	Wtlnds	Pop	Grazing
Kolbe (R)	5	15%	—	—	—	+	—
Kyl (R)	4	0%	—	—	—	—	—
Pastor (D)	2	50%	—	—	+	I	I
Rhodes (R)	1	0%	—	—	—	—	?
Stump (R)	3	0%	—	—	—	—	—

+ = Vote for the environment   * = Senate seat up for election
— = Vote against the environment ? = Absent (counts as anti-environment vote)
Dst = District   I = Ineligible to vote

See pp. 94-96 for complete scorecard key and descriptions of bills.

ARKANSAS

SENATE		LCV '91 SCORE	Energy	Arctic	Wtlnds	Pop	Lands
Bumpers* (D)		47%	—	—	+	+	+
Pryor (D)		27%	—	—	+	?	—

SAMPLE BILLS

HOUSE	Dst	LCV '91 SCORE	Toxic	Arctic	Wtlnds	Pop	Grazing
Alexander (D)	1	38%	—	—	—	+	—
Anthony (D)	4	23%	—	—	—	+	—
Hammerschmidt (R)	3	0%	—	—	—	—	—
Thornton (D)	2	38%	—	—	+	+	?

CALIFORNIA

SENATE		LCV '91 SCORE	Energy	Arctic	Wtlnds	Pop	Lands
Cranston* (D)		80%	+	—	+	+	+
Seymour* (R)		14%	—	—	—	+	—

SAMPLE BILLS

HOUSE	Dst	LCV '91 SCORE	Toxic	Arctic	Wtlnds	Pop	Grazing
Anderson (D)	32	31%	—	—	+	+	+
Beilenson (D)	23	100%	+	+	+	+	+
Berman (D)	26	92%	+	+	+	+	+
Boxer (D)	6	85%	+	+	+	+	+
Brown (D)	36	85%	+	+	+	+	+
Campbell (R)	12	69%	—	—	+	+	+
Condit (D)	15	38%	—	—	—	+	—
Cox (R)	40	23%	—	—	—	—	+
Cunningham (R)	44	0%	—	—	—	—	—
Dannemeyer (R)	39	0%	—	—	—	—	—
Dellums (D)	8	92%	+	+	+	+	+
Dixon (D)	28	85%	+	+	+	+	+
Dooley (D)	17	62%	—	+	+	+	—
Doolittle (R)	14	0%	—	—	—	—	—
Dornan (R)	38	8%	—	—	—	—	—
Dreier (R)	33	31%	—	—	+	—	+
Dymally (D)	31	54%	—	+	+	?	—
Edwards (D)	10	92%	+	+	+	+	+
Fazio (D)	4	69%	+	—	+	+	+
Gallegley (R)	21	0%	—	—	—	—	—

+ = Vote for the environment   * = Senate seat up for election
— = Vote against the environment   ? = Absent (counts as anti-environment vote)
Dst = District

See pp. 94-96 for complete scorecard key and descriptions of bills.

HOUSE	Dst	LCV '91 SCORE	Toxic	Arctic	Wtlnds	Pop	Grazing
Herger (R)	2	0%	—	—	—	—	—
Hunter (R)	45	0%	—	—	—	—	—
Lagomarsino	19	0%	—	—	—	—	—
Lantos (D)	11	85%	+	+	+	+	+
Lehman (D)	18	54%	—	+	+	+	—
Levine (D)	27	69%	+	+	+	+	?
Lewis (R)	35	8%	—	—	—	—	—
Lowery (R)	41	8%	—	—	+	—	—
Martinez (D)	30	62%	+	—	+	+	—
Matsui (D)	3	77%	+	+	+	+	+
McCandless (R)	37	8%	—	—	—	+	—
Miller (D)	7	85%	+	—	+	+	+
Mineta (D)	13	85%	+	+	+	+	+
Moorhead (R)	22	15%	—	—	+	—	—
Packard (R)	43	0%	—	—	—	—	—
Panetta (D)	16	69%	—	+	+	+	—
Pelosi (D)	5	92%	+	+	+	+	+
Riggs (R)	1	31%	—	+	+	+	—
Rohrabacher (R)	42	15%	—	—	+	—	+
Roybal (D)	25	92%	+	+	+	+	+
Stark (D)	9	100%	+	+	+	+	+
Thomas (R)	20	15%	—	—	—	+	—
Torres (D)	34	92%	+	+	+	+	+
Waters (D)	29	77%	+	—	+	+	+
Waxman (D)	24	92%	+	+	+	+	+

COLORADO

SENATE	LCV '91 SCORE	SAMPLE BILLS Energy	Arctic	Wtlnds	Pop	Lands
Brown (R)	36%	—	—	+	+	—
Wirth* (D)	80%	+	+	+	+	—

HOUSE	Dst	LCV '91 SCORE	Toxic	Arctic	Wtlnds	Pop	Grazing
Allard (R)	4	0%	—	—	—	—	—
Campbell (D)	3	38%	+	—	—	+	—
Hefley (R)	5	0%	—	—	—	—	—

+ = Vote for the environment * = Senate seat up for election
— = Vote against the environment ? = Absent (counts as anti-environment vote)
Dst = District

See pp. 94-96 for complete scorecard key and descriptions of bills.

HOUSE	Dst	LCV '91 SCORE	Toxic	Arctic	Wtlnds	Pop	Grazing
Schaefer (R)	6	0%	—	—	—	—	—
Schroeder (D)	1	92%	+	+	+	+	+
Skaggs (D)	2	85%	—	+	+	+	+

CONNECTICUT

SENATE	LCV '91 SCORE	SAMPLE BILLS Energy	Arctic	Wtlnds	Pop	Lands
Dodd* (D)	73%	—	+	+	+	—
Lieberman (D)	100%	+	+	+	+	+

HOUSE	Dst	LCV '91 SCORE	Toxic	Arctic	Wtlnds	Pop	Grazing
DeLauro (D)	3	100%	+	+	+	+	+
Franks (R)	5	15%	—	—	+	+	—
Gejdenson (D)	2	54%	—	—	+	+	+
Johnson (R)	6	38%	—	—	+	+	+
Kennelly (D)	1	77%	—	+	+	+	+
Shays (R)	4	100%	+	+	+	+	+

DELAWARE

SENATE	LCV '91 SCORE	SAMPLE BILLS Energy	Arctic	Wtlnds	Pop	Lands
Biden (D)	93%	+	+	+	+	+
Roth (R)	73%	+	+	+	+	—

HOUSE	Dst	LCV '91 SCORE	Toxic	Arctic	Wtlnds	Pop	Grazing
Carper (D)	1	69%	—	—	+	+	+

FLORIDA

SENATE	LCV '91 SCORE	SAMPLE BILLS Energy	Arctic	Wtlnds	Pop	Lands
Graham* (D)	73%	+	—	+	+	+
Mack (R)	20%	+	—	+	—	?

+ = Vote for the environment * = Senate seat up for election
— = Vote against the environment ? = Absent (counts as anti-environment vote
Dst = District

See pp. 94-96 for complete scorecard key and descriptions of bills.

HOUSE	Dst	LCV '91 SCORE	Toxic	Arctic	Wtlnds	Fund	Grazing
Bacchus (D)	11	77%	+	+	+	+	+
Bennett (D)	3	69%	+	—	+	—	+
Bilirakis (R)	9	15%	—	—	+	—	—
Fascell (D)	19	77%	+	+	+	+	+
Gibbons (D)	7	62%	+	—	+	+	+
Goss (R)	13	46%	—	—	+	—	+
Hutto (D)	1	15%	—	—	—	—	—
Ireland (R)	10	15%	—	—	—	—	+
James (R)	4	46%	—	—	+	—	+
Johnston (D)	14	92%	+	+	+	+	+
Lehman (D)	17	77%	+	+	+	+	+
Lewis (R)	12	23%	—	—	+	—	—
McCollum (R)	5	31%	—	—	+	—	—
Peterson (D)	2	38%	—	—	+	+	—
Ros-Lehtinen (R)	18	54%	—	—	+	—	+
Shaw (R)	15	15%	—	—	+	—	—
Smith (D)	16	77%	+	—	+	+	+
Stearns (R)	6	15%	—	—	+	—	—
Young (R)	8	15%	—	—	+	—	?

GEORGIA

SENATE		LCV '91 SCORE	SAMPLE BILLS Energy	Arctic	Wtlnds	Pop	Lands
Fowler* (D)		60%	+	—	+	+	+
Nunn (D)		40%	—	—	+	+	+

HOUSE	Dst	LCV '91 SCORE	Toxic	Arctic	Wtlnds	Pop	Grazing
Barnard (D)	10	46%	—	—	—	—	+
Darden (D)	7	69%	—	—	+	+	+
Gingrich (R)	6	8%	—	—	—	—	—
Hatcher (D)	2	46%	—	—	+	+	—
Jenkins (D)	9	62%	—	—	+	+	+
Jones (D)	4	92%	+	+	+	+	+
Lewis (D)	5	85%	+	+	+	+	+
Ray (D)	3	38%	—	—	+	—	—
Rowland (D)	8	62%	—	—	+	+	+
Thomas (D)	1	46%	—	—	+	+	—

+ = Vote for the environment   * = Senate seat up for election
— = Vote against the environment   ? = Absent (counts as anti-environment vote)
Dst = District

See pp. 94-96 for complete scorecard key and descriptions of bills.

HAWAII

SENATE

SENATE	LCV '91 SCORE	Energy	Arctic	Wtlnds	Pop	Lands
Akaka (D)	73%	—	+	+	+	+
Inouye* (D)	27%	—	—	—	+	?

HOUSE

HOUSE	Dst	LCV '91 SCORE	Toxic	Arctic	Wtlnds	Pop	Grazing
Abercrombie (D)	1	77%	+	—	+	+	+
Mink (D)	2	92%	+	+	+	+	+

IDAHO

SENATE

SENATE	LCV '91 SCORE	Energy	Arctic	Wtlnds	Pop	Lands
Craig (R)	14%	—	—	—	—	—
Symms* (R)	13%	—	—	—	—	—

HOUSE

HOUSE	Dst	LCV '91 SCORE	Toxic	Arctic	Wtlnds	Pop	Grazing
LaRocco (D)	1	54%	+	—	+	+	—
Stallings (D)	2	31%	+	—	—	—	—

ILLINOIS

SENATE

SENATE	LCV '91 SCORE	Energy	Arctic	Wtlnds	Pop	Lands
Dixon* (D)	53%	+	—	+	+	+
Simon (D)	87%	+	+	+	+	+

HOUSE

HOUSE	Dst	LCV '91 SCORE	Toxic	Arctic	Wtlnds	Pop	Grazing
Annunzio (D)	11	62%	+	+	+	—	+
Bruce (D)	19	38%	—	—	—	—	+
Collins (D)	7	69%	+	—	+	+	+
Costello (D)	21	38%	—	—	—	—	+
Cox (D)	16	85%	+	+	+	+	+
Crane (R)	12	8%	—	—	—	—	—
Durbin (D)	20	92%	+	+	—	+	+
Evans (D)	17	85%	+	+	+	+	+
Ewing (R)	15	0%	—	—	—	I	I
Fawell (R)	13	62%	+	—	+	+	+
Hastert (R)	14	0%	—	—	—	—	—
Hayes (D)	1	77%	+	—	+	+	+

+ = Vote for the environment   * = Senate seat up for election
— = Vote against the environment   ? = Absent (counts as anti-environment vote)
Dst = District   I = Ineligible to vote

See pp. 94-96 for complete scorecard key and descriptions of bills.

HOUSE	Dst	LCV '91 SCORE	Toxic	Arctic	Wtlnds	Pop	Grazing
Hyde (R)	6	8%	—	—	+	—	—
Lipinski (D)	5	62%	+	+	—	—	+
Michel (R)	18	0%	—	—	—	—	—
Porter (R)	10	62%	+	—	+	+	?
Poshard (D)	22	54%	—	+	—	—	+
Rostenkowski (D)	8	54%	—	—	+	?	+
Russo (D)	3	62%	+	+	+	—	+
Sangmeister (D)	4	85%	+	—	+	+	+
Savage (D)	2	69%	+	—	+	+	+
Yates (D)	9	77%	+	+	+	?	+

INDIANA

SENATE	LCV '91 SCORE	SAMPLE BILLS Energy	Arctic	Wtlnds	Pop	Lands
Coats* (R)	20%	—	—	+	—	—
Lugar (R)	27%	—	—	+	—	—

HOUSE	Dst	LCV '91 SCORE	Toxic	Arctic	Wtlnds	Pop	Grazing
Burton (R)	6	0%	—	—	—	—	—
Hamilton (D)	9	62%	—	—	+	+	+
Jacobs (D)	10	92%	+	+	+	+	+
Jontz (D)	5	100%	+	+	+	+	+
Long (D)	4	46%	—	—	+	+	—
McCloskey (D)	8	69%	—	—	+	+	+
Myers (R)	7	0%	—	—	—	—	—
Roemer (D)	3	62%	—	+	+	—	+
Sharp (D)	2	62%	—	—	+	+	+
Visclosky (D)	1	54%	—	—	+	+	+

IOWA

SENATE	LCV '91 SCORE	SAMPLE BILLS Energy	Arctic	Wtlnds	Pop	Lands
Grassley* (R)	33%	+	—	+	—	—
Harkin (D)	73%	+	+	+	?	—

+ = Vote for the environment   * = Senate seat up for election
— = Vote against the environment   ? = Absent (counts as anti-environment vote)
Dst = District

See pp. 94-96 for complete scorecard key and descriptions of bills.

103

HOUSE	Dst	LCV '91 SCORE	Toxic	Arctic	Wtlnds	Pop	Grazing
Grandy (R)	6	8%	—	—	—	—	—
Leach (R)	1	46%	—	—	—	+	—
Lightfoot (R)	5	0%	—	—	—	—	—
Nagle (D)	3	77%	—	+	+	+	—
Nussle (R)	2	8%	—	—	—	—	—
Smith (D)	4	38%	—	—	+	+	—

KANSAS

SENATE	LCV '91 SCORE	SAMPLE BILLS				
		Energy	Arctic	Wtlnds	Pop	Lands
Dole* (R)	13%	—	—	+	—	—
Kassebaum (R)	40%	—	—	+	+	—

HOUSE	Dst	LCV '91 SCORE	Toxic	Arctic	Wtlnds	Pop	Grazing
Glickman (D)	4	54%	—	—	+	+	+
Meyers (R)	3	54%	—	—	+	+	+
Nichols (R)	5	8%	—	—	—	+	—
Roberts (R)	1	0%	—	—	—	—	—
Slattery (D)	2	54%	—	+	—	+	+

KENTUCKY

SENATE	LCV '91 SCORE	SAMPLE BILLS				
		Energy	Arctic	Wtlnds	Pop	Lands
Ford* (D)	7%	—	—	+	—	—
McConnell (R)	13%	—	—	—	—	—

HOUSE	Dst	LCV '91 SCORE	Toxic	Arctic	Wtlnds	Pop	Grazing
Bunning (R)	4	8%	—	—	—	—	—
Hopkins (R)	6	8%	—	—	+	?	—
Hubbard (D)	1	38%	—	—	—	+	—
Mazzoli (D)	3	46%	—	—	—	—	+
Natcher (D)	2	46%	—	—	+	—	—
Perkins (D)	7	38%	—	—	+	—	—
Rogers (R)	5	0%	—	—	—	—	—

+ = Vote for the environment   * = Senate seat up for election
— = Vote against the environment   ? = Absent (counts as anti-environment vote)
Dst = District

See pp. 94-96 for complete scorecard key and descriptions of bills.

LOUISIANA

SENATE		LCV '91 SCORE	Energy	Arctic	Wtlnds	Pop	Lands
					SAMPLE BILLS		
Breaux* (D)		20%	—	—	—	—	—
Johnston (D)		13%	—	—	—	—	—

HOUSE	Dst	LCV '91 SCORE	Toxic	Arctic	Wtlnds	Pop	Grazing
Baker (R)	6	0%	—	—	—	—	—
Hayes (D)	7	23%	—	—	—	—	+
Holloway (R)	8	0%	—	—	—	—	—
Huckaby (D)	5	38%	—	—	—	—	+
Jefferson (D)	2	62%	+	—	—	+	+
Livingston (R)	1	0%	—	—	—	—	—
McCrery (R)	4	8%	—	—	—	—	+
Tauzin (D)	3	15%	—	—	—	—	+

MAINE

SENATE		LCV '91 SCORE	Energy	Arctic	Wtlnds	Pop	Lands
					SAMPLE BILLS		
Cohen (R)		93%	+	—	+	+	+
Mitchell (D)		80%	—	—	+	+	+

HOUSE	Dst	LCV '91 SCORE	Toxic	Arctic	Wtlnds	Pop	Grazing
Andrews (D)	1	100%	+	+	+	+	+
Snowe (R)	2	62%	+	—	+	+	+

MARYLAND

SENATE		LCV '91 SCORE	Energy	Arctic	Wtlnds	Pop	Lands
					SAMPLE BILLS		
Mikulski* (D)		93%	+	+	+	+	+
Sarbanes (D)		87%	+	—	+	+	+

HOUSE	Dst	LCV '91 SCORE	Toxic	Arctic	Wtlnds	Pop	Grazing
Bentley (R)	2	8%	—	—	—	+	—
Byron (D)	6	38%	—	—	+	—	+
Cardin (D)	3	62%	+	—	+	+	+
Gilchrest (R)	1	54%	+	—	+	+	—

+ = Vote for the environment   * = Senate seat up for election
— = Vote against the environment
Dst = District

See pp. 94-96 for complete scorecard key and descriptions of bills.

HOUSE	Dst	LCV '91 SCORE	Toxic	Arctic	Wtlnds	Pop	Grazing
Hoyer (D)	5	69%	+	—	+	+	+
McMillen (D)	4	77%	+	+	+	+	+
Mfume (D)	7	85%	+	+	+	+	+
Morella (R)	8	92%	+	+	+	+	+

MASSACHUSETTS

SENATE	LCV '91 SCORE	SAMPLE BILLS Energy	Arctic	Wtlnds	Pop	Lands
Kennedy (D)	100%	+	+	+	+	+
Kerry (D)	93%	+	+	+	+	+

HOUSE	Dst	LCV '91 SCORE	Toxic	Arctic	Wtlnds	Pop	Grazing
Atkins (D)	5	85%	+	+	+	+	+
Donnelly (D)	11	69%	+	—	+	—	+
Early (D)	3	54%	—	—	+	—	+
Frank (D)	4	100%	+	+	+	+	+
Kennedy (D)	8	92%	+	+	+	+	+
Markey (D)	7	92%	+	—	+	+	+
Mavroules (D)	6	69%	+	—	+	—	+
Moakley (D)	9	54%	—	—	+	—	+
Neal (D)	2	69%	+	+	+	—	+
Olver (D)	1	89%	+	+	+	I	+
Studds (D)	10	85%	+	—	+	+	+

MICHIGAN

SENATE	LCV '91 SCORE	SAMPLE BILLS Energy	Arctic	Wtlnds	Pop	Lands
Levin (D)	73%	+	—	+	+	+
Riegle (D)	67%	+	—	+	+	+

HOUSE	Dst	LCV '91 SCORE	Toxic	Arctic	Wtlnds	Pop	Grazing
Bonior (D)	12	92%	+	+	+	+	+
Broomfield (R)	18	15%	—	—	—	—	—
Camp (R)	10	15%	—	—	—	—	—
Carr (D)	6	46%	—	—	—	+	—

+ = Vote for the environment   * = Senate seat up for election
— = Vote against the environment   I = Ineligible to vote
Dst = District

See pp. 94-96 for complete scorecard key and descriptions of bills.

HOUSE	Dst	LCV '91 SCORE	Toxic	Arctic	Wtlnds	Pop	Grazing
Collins (D)	13	62%	—	—	+	+	—
Conyers (D)	1	92%	+	+	+	+	+
Davis (R)	11	15%	—	—	+	?	—
Dingell (D)	16	46%	—	—	+	?	+
Ford (D)	15	69%	—	—	+	+	+
Henry (R)	5	54%	—	—	+	—	+
Hertel (D)	14	77%	+	—	+	+	+
Kildee (D)	7	77%	+	+	+	—	+
Levin (D)	17	77%	+	—	+	+	+
Pursell (R)	2	31%	—	—	+	—	?
Traxler (D)	8	46%	—	—	+	—	+
Upton (R)	4	54%	—	—	+	+	+
Vander Jagt (R)	9	8%	—	—	—	—	—
Wolpe (D)	3	100%	+	+	+	+	+

MINNESOTA

SENATE	LCV '91 SCORE	SAMPLE BILLS				
		Energy	Arctic	Wtlnds	Pop	Lands
Durenberger (R)	60%	+	+	+	—	—
Wellstone (D)	93%	+	+	+	+	+

HOUSE	Dst	LCV '91 SCORE	Toxic	Arctic	Wtlnds	Pop	Grazing
Oberstar (D)	8	54%	—	+	+	—	—
Penny (D)	1	62%	+	+	+	+	—
Peterson (D)	7	54%	+	+	+	—	—
Ramstad (R)	3	62%	—	+	+	—	+
Sabo (D)	5	77%	—	+	+	+	+
Sikorski (D)	6	92%	+	+	+	+	+
Vento (D)	4	69%	+	+	+	+	+
Weber (R)	2	15%	—	+	—	—	—

MISSISSIPPI

SENATE	LCV '91 SCORE	SAMPLE BILLS				
		Energy	Arctic	Wtlnds	Pop	Lands
Cochran (R)	7%	—	—	—	?	—
Lott (R)	20%	+	—	—	—	—

+ = Vote for the environment * = Senate seat up for election
— = Vote against the environment ? = Absent (counts as anti-environment vote)
Dst = District

See pp. 94-96 for complete scorecard key and descriptions of bills.

HOUSE	Dst	LCV '91 SCORE	Toxic	Arctic	Wtlnds	Pop	Grazing
Espy (D)	2	54%	—	—	—	+	—
Montgomery (D)	3	23%	—	—	—	—	—
Parker (D)	4	23%	—	—	—	—	—
Taylor (D)	5	15%	—	—	+	—	—
Whitten (D)	1	46%	—	—	+	—	—

MISSOURI

SENATE		LCV '91 SCORE	SAMPLE BILLS				
			Energy	Arctic	Wtlnds	Pop	Lands
Bond* (R)		13%	+	—	—	—	—
Danforth (R)		27%	—	—	+	—	—

HOUSE	Dst	LCV '91 SCORE	Toxic	Arctic	Wtlnds	Pop	Grazing
Clay (D)	1	85%	+	+	+	+	+
Coleman (R)	6	0%	—	—	—	—	—
Emerson (R)	8	0%	—	—	—	—	—
Gephardt (D)	3	54%	—	—	+	+	+
Hancock (R)	7	0%	—	—	—	—	—
Horn (D)	2	77%	—	+	+	+	+
Skelton (D)	4	23%	—	—	—	—	—
Volkmer (D)	9	23%	—	—	—	—	—
Wheat (D)	5	85%	+	+	+	+	+

MONTANA

SENATE		LCV '91 SCORE	SAMPLE BILLS				
			Energy	Arctic	Wtlnds	Pop	Lands
Baucus (D)		80%	+	+	+	+	—
Burns (R)		13%	—	—	—	—	—

HOUSE	Dst	LCV '91 SCORE	Toxic	Arctic	Wtlnds	Pop	Grazing
Marlenee (R)	2	0%	—	—	—	—	—
Williams (D)	1	54%	—	+	+	+	—

+ = Vote for the environment   * = Senate seat up for election
— = Vote against the environment
Dst = District

See pp. 94-96 for complete scorecard key and descriptions of bills.

NEBRASKA

SENATE

SENATE	LCV '91 SCORE	Energy	Arctic	Wtlnds	Pop	Lands
Exon (D)	53%	+	—	+	+	—
Kerrey (D)	80%	?	—	+	+	—

SAMPLE BILLS

HOUSE

HOUSE	Dst	LCV '91 SCORE	Toxic	Arctic	Wtlnds	Pop	Grazing
Barrett (R)	3	0%	—	—	—	—	—
Bereuter (R)	1	46%	—	—	+	—	—
Hoagland (D)	2	85%	—	—	+	+	+

NEVADA

SENATE

SENATE	LCV '91 SCORE	Energy	Arctic	Wtlnds	Pop	Lands
Bryan (D)	67%	+	+	+	+	—
Reid* (D)	53%	+	—	+	—	—

SAMPLE BILLS

HOUSE

HOUSE	Dst	LCV '91 SCORE	Toxic	Arctic	Wtlnds	Pop	Grazing
Bilbray (D)	1	54%	+	—	+	—	—
Vucanovich (R)	2	0%	—	—	—	—	—

NEW HAMPSHIRE

SENATE

SENATE	LCV '91 SCORE	Energy	Arctic	Wtlnds	Pop	Lands
Rudman* (R)	53%	—	—	+	+	+
Smith (R)	50%	+	—	+	—	+

SAMPLE BILLS

HOUSE

HOUSE	Dst	LCV '91 SCORE	Toxic	Arctic	Wtlnds	Pop	Grazing
Swett (D)	2	77%	+	+	+	—	+
Zeliff (R)	1	23%	—	—	—	+	+

NEW JERSEY

SENATE

SENATE	LCV '91 SCORE	Energy	Arctic	Wtlnds	Pop	Lands
Bradley (D)	87%	+	+	+	+	+
Launtenberg (D)	100%	+	+	+	+	+

SAMPLE BILLS

+ = Vote for the environment   * = Senate seat up for election
— = Vote against the environment   ? = Absent (counts as anti-environment vote)
Dst = District

See pp. 94-96 for complete scorecard key and descriptions of bills.

HOUSE	Dst	LCV '91 SCORE	Toxic	Arctic	Wtlnds	Pop	Grazing
Andrews (D)	1	62%	+	+	+	+	—
Dwyer (D)	6	92%	+	+	+	+	+
Gallo (R)	11	38%	—	—	+	+	+
Guarini (D)	14	92%	+	+	+	+	+
Hughes (D)	2	92%	+	—	+	+	+
Pallone (D)	3	85%	+	+	+	+	+
Payne (D)	10	92%	+	—	+	+	+
Rinaldo (R)	7	62%	—	+	+	—	+
Roe (D)	8	31%	—	—	+	—	—
Roukema (R)	5	54%	—	—	+	?	+
Saxton (R)	13	54%	—	—	+	—	+
Smith (R)	4	62%	+	—	+	—	+
Torricelli (D)	9	69%	+	—	+	+	+
Zimmer (R)	12	85%	—	+	+	+	+

NEW MEXICO

SENATE		LCV '91 SCORE	Energy	Arctic	Wtlnds	Pop	Lands
Bingaman (D)		40%	—	—	+	+	—
Domenici (R)		13%	—	—	+	—	—

HOUSE	Dst	LCV '91 SCORE	Toxic	Arctic	Wtlnds	Pop	Grazing
Richardson (D)	3	69%	+	—	+	+	—
Schiff (R)	1	23%	—	—	+	+	—
Skeen (R)	2	0%	—	—	—	—	—

NEW YORK

SENATE		LCV '91 SCORE	Energy	Arctic	Wtlnds	Pop	Lands
D'Amato* (R)		40%	—	—	+	+	—
Moynihan (D)		93%	+	+	+	+	+

HOUSE	Dst	LCV '91 SCORE	Toxic	Arctic	Wtlnds	Pop	Grazing
Ackerman (D)	7	77%	+	+	+	+	+
Boehlert (R)	25	77%	+	+	+	+	+
Downey (D)	2	77%	+	+	+	+	+
Engel (D)	19	85%	—	+	+	+	+

+ = Vote for the environment
— = Vote against the environment
Dst = District
* = Senate seat up for election
? = Absent (counts as anti-environment vote)

See pp. 94-96 for complete scorecard key and descriptions of bills.

HOUSE	Dst	LCV '91 SCORE	Toxic	Arctic	Wtlnds	Pop	Grazing
Fish (R)	21	85%	+	+	+	+	+
Flake (D)	6	69%	—	+	+	+	+
Gilman (R)	22	54%	—	—	—	+	+
Green (R)	15	77%	—	+	+	+	+
Hochbrueckner (D)	1	62%	+	—	+	+	+
Horton (R)	29	54%	+	—	—	+	—
Houghton (R)	34	8%	—	—	—	+	—
LaFalce (D)	32	46%	—	—	—	—	+
Lent (R)	4	8%	—	—	—	—	—
Lowey (D)	20	85%	+	—	+	+	+
Manton (D)	9	62%	—	—	+	—	+
Martin (R)	26	0%	—	—	—	?	—
McGrath (R)	5	46%	—	—	+	—	+
McHugh (D)	28	62%	—	—	+	+	+
McNulty (D)	23	54%	—	—	—	+	+
Molinari (R)	14	38%	+	—	+	+	—
Mrazek (D)	3	54%	+	+	+	+	+
Nowak (D)	33	54%	—	—	+	—	+
Owens (D)	12	77%	+	+	+	+	+
Paxon (R)	31	15%	—	—	—	—	—
Rangel (D)	16	77%	—	+	+	+	+
Scheuer (D)	8	92%	+	+	+	+	+
Schumer (D)	10	92%	+	+	+	+	+
Serrano (D)	18	85%	+	+	+	+	+
Slaughter (D)	30	92%	+	+	+	+	+
Solarz (D)	13	77%	+	—	+	+	+
Solomon (R)	24	8%	—	—	—	—	+
Towns (D)	11	85%	+	+	+	+	+
Walsh (R)	27	31%	—	—	+	—	—
Weiss (D)	17	85%	+	+	+	+	+

NORTH CAROLINA

SENATE	LCV '91 SCORE	SAMPLE BILLS Energy	Arctic	Wtlnds	Pop	Lands
Helms (R)	7%	—	—	—	—	—
Sanford* (D)	67%	+	—	+	+	—

+ = Vote for the environment   * = Senate seat up for election
— = Vote against the environment   ? = Absent (counts as anti-environment vote)
Dst = District

See pp. 94-96 for complete scorecard key and descriptions of bills.

HOUSE	Dst	LCV '91 SCORE	Toxic	Arctic	Wtlnds	Pop	Grazing
Ballenger (R)	10	0%	—	—	—	—	—
Coble (R)	6	15%	—	—	—	—	—
Hefner (D)	8	54%	—	—	—	+	+
Jones (D)	1	31%	—	—	—	+	—
Lancaster (D)	3	69%	—	+	—	+	+
McMillan (R)	9	15%	—	—	—	—	—
Neal (D)	5	100%	+	+	+	+	+
Price (D)	4	92%	+	+	+	+	+
Rose (D)	7	54%	—	—	—	+	—
Taylor (R)	11	0%	—	—	—	—	—
Valentine (D)	2	69%	—	+	—	+	+

NORTH DAKOTA

SENATE		LCV '91 SCORE	SAMPLE BILLS Energy	Arctic	Wtlnds	Pop	Lands
Burdick (D)		60%	+	—	+	+	—
Conrad* (D)		47%	—	—	+	+	—

HOUSE	Dst	LCV '91 SCORE	Toxic	Arctic	Wtlnds	Pop	Grazing
Dorgan (D)	1	15%	+	—	—	—	—

OHIO

SENATE		LCV '91 SCORE	SAMPLE BILLS Energy	Arctic	Wtlnds	Pop	Lands
Glenn* (D)		73%	+	—	+	+	+
Metzenbaum (D)		93%	+	+	+	+	+

HOUSE	Dst	LCV '91 SCORE	Toxic	Arctic	Wtlnds	Pop	Grazing
Applegate (D)	18	31%	—	—	—	—	+
Boehner (R)	8	0%	—	—	—	—	—
Eckart (D)	11	77%	+	—	+	+	+
Feighan (D)	19	77%	+	—	+	+	+
Gillmor (R)	5	23%	—	—	—	—	—
Gradison (R)	2	23%	—	—	+	—	+
Hall (D)	3	46%	—	—	+	—	+
Hobson (R)	7	8%	—	—	—	+	—

+ = Vote for the environment   * = Senate seat up for election
— = Vote against the environment
Dst = District

See pp. 94-96 for complete scorecard key and descriptions of bills.

HOUSE	Dst	LCV '91 SCORE	Toxic	Arctic	Wtlnds	Pop	Grazing
Kaptur (D)	9	62%	—	—	+	+	+
Kasich (R)	12	23%	—	—	+	—	+
Luken (D)	1	62%	—	—	+	—	+
McEwen (R)	6	0%	—	—	—	—	—
Miller (R)	10	8%	—	—	—	—	+
Oakar (D)	20	54%	—	—	+	+	+
Oxley (R)	4	0%	—	—	—	—	—
Pease (D)	13	77%	+	—	—	+	+
Regula (R)	16	8%	—	—	—	—	—
Sawyer (D)	14	62%	+	—	+	+	+
Stokes (D)	21	77%	+	+	+	+	+
Traficant (D)	17	62%	+	—	—	+	—
Wylie (R)	15	8%	—	—	—	—	—

OKLAHOMA

SENATE	LCV '91 SCORE	SAMPLE BILLS				
		Energy	Arctic	Wtlnds	Pop	Lands
Boren (D)	33%	?	—	+	—	—
Nickles* (R)	7%	—	—	—	—	—

HOUSE	Dst	LCV '91 SCORE	Toxic	Arctic	Wtlnds	Pop	Grazing
Brewster (D)	3	23%	—	—	—	+	—
Edwards (R)	5	0%	—	—	—	—	—
English (D)	6	31%	—	—	+	—	—
Inhofe (R)	1	0%	—	—	—	—	—
McCurdy (D)	4	54%	—	—	+	+	+
Synar (D)	2	77%	+	—	+	+	+

OREGON

SENATE	LCV '91 SCORE	SAMPLE BILLS				
		Energy	Arctic	Wtlnds	Pop	Lands
Hatfield (R)	27%	—	—	+	+	—
Packwood* (R)	13%	—	—	—	+	—

HOUSE	Dst	LCV '91 SCORE	Toxic	Arctic	Wtlnds	Pop	Grazing
AuCoin (D)	1	77%	+	+	+	+	—
DeFazio (D)	4	77%	+	+	+	+	—

+ = Vote for the environment   * = Senate seat up for election
— = Vote against the environment   ? = Absent (counts as anti-environment vote)
Dst = District

See pp. 94-96 for complete scorecard key and descriptions of bills.

HOUSE	Dst	LCV '91 SCORE	Toxic	Arctic	Wtlnds	Pop	Grazing
Kopetski (D)	5	54%	—	—	+	+	—
Smith (R)	2	0%	—	—	—	—	—
Wyden (D)	3	77%	+	—	+	+	+

PENNSYLVANIA

SENATE	LCV '91 SCORE	SAMPLE BILLS				
		Energy	Arctic	Wtlnds	Pop	Lands
Specter* (R)	40%	—	—	+	+	—
Wofford (D)	86%	?	—	+	+	+

HOUSE	Dst	LCV '91 SCORE	Toxic	Arctic	Wtlnds	Pop	Grazing
Blackwell (D)	2	57%	—	—	+	I	I
Borski (D)	3	46%	—	—	+	—	+
Clinger (R)	23	15%	—	—	—	—	+
Coughlin (R)	13	46%	—	—	+	+	—
Coyne (D)	14	77%	+	+	+	+	+
Foglietta (D)	1	62%	—	—	+	+	+
Gaydos (D)	20	38%	+	—	+	—	—
Gekas (R)	17	8%	—	—	—	+	—
Goodling (R)	19	0%	—	—	—	—	—
Gray (D)	2	40%	—	—	+	+	?
Kanjorski (D)	11	62%	—	—	+	—	+
Kolter (D)	4	31%	—	—	—	?	?
Kostmayer (D)	8	92%	+	+	+	+	+
McDade (R)	10	23%	—	—	—	—	—
Murphy (D)	22	46%	+	—	—	—	+
Murtha (D)	12	54%	—	—	+	—	—
Ridge (R)	21	8%	—	—	—	—	—
Ritter (R)	15	31%	—	—	—	—	+
Santorum (R)	18	0%	—	—	—	—	—
Schulze (R)	5	8%	—	—	+	—	—
Shuster (R)	9	0%	—	—	—	—	—
Walker (R)	16	8%	—	—	—	—	+
Weldon (R)	7	46%	—	—	+	—	+
Yatron (D)	6	38%	—	—	+	—	+

+ = Vote for the environment
— = Vote against the environment
Dst = District
* = Senate seat up for election
? = Absent (counts as anti-environment vote)
I = Ineligible to vote

See pp. 94-96 for complete scorecard key and descriptions of bills.

RHODE ISLAND

| | | LCV '91 | SAMPLE BILLS | | | | |
SENATE		SCORE	Energy	Arctic	Wtlnds	Pop	Lands
Chafee (R)		80%	+	+	+	+	+
Pell (D)		100%	P	+	+	+	+

| | | | LCV '91 | SAMPLE BILLS | | | | |
HOUSE	Dst		SCORE	Toxic	Arctic	Wtlnds	Pop	Grazing
Machtley (R)	1		77%	+	—	+	+	+
Reed (D)	2		92%	+	+	+	+	—

SOUTH CAROLINA

| | | LCV '91 | SAMPLE BILLS | | | | |
SENATE		SCORE	Energy	Arctic	Wtlnds	Pop	Lands
Hollings* (D)		67%	+	—	+	+	+
Thurmond (R)		7%	—	—	—	—	—

| | | LCV '91 | SAMPLE BILLS | | | | |
HOUSE	Dst	SCORE	Toxic	Arctic	Wtlnds	Pop	Grazing
Derrick (D)	3	77%	—	—	+	+	+
Patterson (D)	4	85%	—	+	+	+	+
Ravenel (R)	1	100%	+	+	+	+	+
Spence (R)	2	8%	—	—	—	—	—
Spratt (D)	5	92%	+	+	+	+	+
Tallon (D)	6	38%	—	—	—	—	—

SOUTH DAKOTA

| | | LCV '91 | SAMPLE BILLS | | | | |
SENATE		SCORE	Energy	Arctic	Wtlnds	Pop	Lands
Daschle* (D)		53%	—	—	+	+	—
Pressler (R)		20%	—	—	+	—	—

| | | LCV '91 | | | | | |
HOUSE	Dst	SCORE	Toxic	Arctic	Wtlnds	Pop	Grazing
Johnson (D)	1	69%	+	+	+	+	—

+ = Vote for the environment   * = Senate seat up for election
— = Vote against the environment   P = Present but didn't vote
Dst = District

See pp. 94-96 for complete scorecard key and descriptions of bills.

TENNESSEE

SENATE		LCV '91 SCORE	SAMPLE BILLS				
			Energy	Arctic	Wtlnds	Pop	Lands
Gore (D)		73%	+	+	+	+	+
Sasser (D)		53%	+	—	+	+	+

HOUSE	Dst	LCV '91 SCORE	Toxic	Arctic	Wtlnds	Pop	Grazing
Clement (D)	5	54%	—	—	—	+	+
Cooper (D)	4	77%	—	+	+	+	+
Duncan (R)	2	8%	—	—	—	—	—
Ford (D)	9	77%	+	—	+	+	+
Gordon (D)	6	54%	—	—	—	+	+
Lloyd (D)	3	38%	—	—	+	+	+
Quillen (R)	1	0%	—	—	—	—	—
Sundquist (R)	7	8%	—	—	—	—	—
Tanner (D)	8	31%	—	—	—	+	+

TEXAS

SENATE		LCV '91 SCORE	SAMPLE BILLS				
			Energy	Arctic	Wtlnds	Pop	Lands
Bentsen (D)		40%	—	—	—	?	—
Gramm (R)		20%	?	—	—	—	—

HOUSE	Dst	LCV '91 SCORE	Toxic	Arctic	Wtlnds	Pop	Grazing
Andrews (D)	25	69%	—	—	+	+	+
Archer (R)	7	23%	—	—	+	—	+
Armey (R)	26	0%	—	—	—	—	—
Barton (R)	6	0%	—	—	—	—	—
Brooks (D)	9	38%	—	—	—	+	+
Bryant (D)	5	77%	+	—	+	+	+
Bustamante (D)	23	62%	+	—	+	+	—
Chapman (D)	1	0%	—	—	—	—	—
Coleman (D)	16	62%	—	+	+	+	—
Combest (R)	19	0%	—	—	—	—	—
de la Garza (D)	15	31%	—	—	+	—	—
DeLay (R)	22	8%	—	—	—	—	+
Edwards (D)	11	38%	—	—	—	+	—
Fields (R)	8	0%	—	—	—	—	—
Frost (D)	24	54%	+	—	+	+	—
Geren (D)	12	31%	—	—	—	+	—

+ = Vote for the environment   * = Senate seat up for election
— = Vote against the environment   ? = Absent (counts as anti-environment vote)
Dst = District

See pp. 94-96 for complete scorecard key and descriptions of bills.

HOUSE	Dst	LCV '91 SCORE	Toxic	Arctic	Wtlnds	Pop	Grazing
Gonzalez (D)	20	54%	—	+	+	+	—
Hall (D)	4	0%	—	—	—	—	—
Johnson (R)	3	9%	—	—	+	—	—
Laughlin (D)	14	15%	—	—	—	—	—
Ortiz (D)	27	23%	—	—	—	—	—
Pickle (D)	10	38%	—	—	+	+	—
Sarpalius (D)	13	23%	—	—	—	—	—
Smith (R)	21	23%	—	—	—	+	+
Stenholm (D)	17	8%	—	—	—	—	—
Washington (D)	18	69%	—	+	+	+	+
Wilson (D)	2	46%	—	—	—	+	+

UTAH

SENATE	LCV '91 SCORE	SAMPLE BILLS Energy	Arctic	Wtlnds	Pop	Lands
Garn* (R)	13%	—	—	—	—	—
Hatch (R)	13%	—	—	—	—	—

HOUSE	Dst	LCV '91 SCORE	Toxic	Arctic	Wtlnds	Pop	Grazing
Hansen (R)	1	0%	—	—	—	—	—
Orton (D)	3	15%	—	—	—	—	—
Owens (D)	2	77%	—	+	+	+	—

VERMONT

SENATE	LCV '91 SCORE	SAMPLE BILLS Energy	Arctic	Wtlnds	Pop	Lands
Jeffords (R)	87%	?	+	+	+	+
Leahy* (D)	100%	+	+	+	+	+

HOUSE	Dst	LCV '91 SCORE	Toxic	Arctic	Wtlnds	Pop	Grazing
Sanders (Ind.)	1	85%	+	+	+	+	+

VIRGINIA

SENATE	LCV '91 SCORE	SAMPLE BILLS Energy	Arctic	Wtlnds	Pop	Lands
Robb (D)	80%	+	—	+	+	+
Warner (R)	27%	—	—	+	—	+

+ = Vote for the environment   * = Senate seat up for election
— = Vote against the environment   ? = Absent (counts as anti-environment vote)
Dst = District

See pp. 94-96 for complete scorecard key and descriptions of bills.

HOUSE	Dst	LCV '91 SCORE	Toxic	Arctic	Wtlnds	Pop	Grazing
Allen (R)	7	14%	—	—	+	I	I
Bateman (R)	1	8%	—	—	—	—	—
Bliley (R)	3	0%	—	—	—	—	—
Boucher (D)	9	62%	—	—	—	+	+
Moran (D)	8	54%	+	+	+	+	—
Olin (D)	6	38%	—	+	—	+	—
Payne (D)	5	69%	+	—	—	+	+
Pickett (D)	2	38%	—	—	—	+	+
Sisisky (D)	4	38%	—	—	—	+	+
Slaughter (R)	7	0%	—	—	—	—	—
Wolf (R)	10	15%	—	—	+	—	—

WASHINGTON

SENATE	LCV '91 SCORE	SAMPLE BILLS				
		Energy	Arctic	Wtlnds	Pop	Lands
Adams* (D)	87%	+	+	+	+	—
Gorton (R)	40%	—	—	+	—	—

HOUSE	Dst	LCV '91 SCORE	Toxic	Arctic	Wtlnds	Pop	Grazing
Chandler (R)	8	15%	—	—	—	+	—
Dicks (D)	6	54%	—	—	+	+	+
Foley (D)	5	Speaker of the House only votes to break a tie					
McDermott (D)	7	85%	+	—	+	+	+
Miller (R)	1	46%	—	—	+	+	+
Morrison (R)	4	31%	—	—	—	+	—
Swift (D)	2	46%	—	—	—	+	—
Unsoeld (D)	3	85%	+	—	+	+	+

WEST VIRGINIA

SENATE	LCV '91 SCORE	SAMPLE BILLS				
		Energy	Arctic	Wtlnds	Pop	Lands
Byrd (D)	33%	—	—	+	+	—
Rockefeller (D)	80%	+	—	+	+	—

+ = Vote for the environment
— = Vote against the environment
Dst = District
* = Senate seat up for election
I = Ineligible to vote

See pp. 94-96 for complete scorecard key and descriptions of bills.

HOUSE	Dst	LCV '91 SCORE	Toxic	Arctic	Wtlnds	Pop	Grazing
Mollohan (D)	1	38%	—	—	+	—	—
Rahall (D)	4	54%	+	—	+	—	+
Staggers (D)	2	46%	—	—	+	—	—
Wise (D)	3	54%	—	—	+	+	+

WISCONSIN

SENATE		LCV '91 SCORE	SAMPLE BILLS				
			Energy	Arctic	Wtlnds	Pop	Lands
Kasten* (R)		40%	—	—	+	—	+
Kohl (D)		73%	+	—	+	+	+

HOUSE	Dst	LCV '91 SCORE	Toxic	Arctic	Wtlnds	Pop	Grazing
Aspin (D)	1	62%	—	+	+	+	+
Gunderson (R)	3	23%	—	—	—	—	—
Kleczka (D)	4	69%	—	—	+	+	+
Klug (R)	2	31%	—	—	+	+	—
Moody (D)	5	92%	+	+	+	+	+
Obey (D)	7	58%	—	—	+	—	+
Petri (R)	6	38%	—	—	+	—	+
Roth (R)	8	23%	—	—	—	—	—
Sensenbrenner (R)	9	23%	—	—	+	—	+

WYOMING

SENATE		LCV '91 SCORE	SAMPLE BILLS				
			Energy	Arctic	Wtlnds	Pop	Lands
Simpson (R)		13%	—	—	—	+	—
Wallop (R)		7%	—	—	—	—	—

HOUSE	Dst	LCV '91 SCORE	Toxic	Arctic	Wtlnds	Pop	Grazing
Thomas (R)	1	0%	—	—	—	—	—

+ = Vote for the environment * = Senate seat up for election
— = Vote against the environment
Dst = District

See pp. 94-96 for complete scorecard key and descriptions of bills.

FIND OUT WHAT'S GOING ON IN YOUR STATE

Throughout this book, we've provided ways to evaluate the environmental records of candidates at the federal, state, and local levels.
If you'd like more details about candidates running for office in your state, some of the groups listed in this section may be able to help. Many of them endorse candidates and provide legislative summaries; most can answer questions about important state and local issues.

ALABAMA

The Sierra Club, P.O. Box 55591, Birmingham, AL 35255. Phone: (205) 933-9269.
Makes endorsements in state legislative races.

ALASKA

Alaska Environmental Lobby, P.O. Box 22151, Juneau, AK 99802. Phone: (907) 463-3366.
A coalition of 19 state groups. Publishes a legislative voting chart.

Alaska Sierra Club, 241 E. 5th Ave., Suite 205, Anchorage, AK 99501. Phone: (907) 276-4048.
Makes endorsements in state legislative races.

ARIZONA

Grand Canyon Chapter Sierra Club, 516 E. Portland St., Phoenix, AZ 85004. Phone: (602) 253-8633.
Publishes a legislative scorecard and makes endorsements in state races.

ARKANSAS

Arkansans for Environmental Reform, P.O. Box 2494, Little Rock, AR 72203. Phone: (501) 397-5576.
Lobbies the legislature on environmental issues, makes endorsements in state races.

Arkansas Sierra Club, Political Chair Mike Faupel, 600 W. Cherry St., Fayetteville, AR 85004.
Publishes a legislative scorecard, makes endorsements in state races.

CALIFORNIA

California League of Conservation Voters, 965 Mission, Suite 750, San Francisco, CA 94103. Phone: (415) 896-5550.
Publishes an annual "report card" on state legislators and endorses state candidates.

CalPIRG, 11965 Venice Blvd. #408, Los Angeles, CA 90066-3954. Phone: (310) 397-3404.
Publishing a legislative scorecard for the November 1992 elections.

COLORADO

Sierra Club, Rocky Mountain Chapter, 777 Grant St., Suite 606, Denver, CO 80203. Phone: (303) 861-8819.
Produces a scorecard that rates legislators and endorses candidates in state legislative races.

CONNECTICUT

Sierra Club, 118 Oak Street, Hartford, CT 06106. Phone: (203) 527-9788.
Publishes an endorsement sheet on state candidates.

DELAWARE

Delaware Nature Society, P.O. Box 700, Hockessin, DE 19707. Phone: (302) 239-2334.
General information and advice.

FLORIDA

League of Conservation Voters, P.O. Box 857, Floral City, FL 32636.
Publishes an in-depth environmental scorecard on local legislators.

GEORGIA

Sierra Club, P.O. Box 467151, Atlanta, GA 30346. Phone: (404) 607-1262.
Publishes an endorsement sheet for state candidates.

HAWAII

Honolulu League of Conservation Voters, P.O. Box 27705, Honolulu, HI 96827.
Endorses local legislators.

IDAHO

Idaho Conservation League, P.O. Box 844, Boise, ID 83701.
Phone: (208) 345-6933.
Their 1992 Legislative Voting Record (in the May issue of their newsletter) describes 7 key environmental bills and shows how legislators voted on them.

Idaho Rivers United, P.O. Box 633, Boise, ID 83701.
Phone: (208) 343-7481.
Publishes election scorecards.

ILLINOIS

Illinois Environmental Council, 313 W. Cook, Springfield, IL 62704. Phone: (217) 544-5954.
Publishes an Environmental Voting Record.

Sierra Club, 506 S. Wabash St. # 505, Chicago, IL 60605.
Phone: (312) 431-0158.
Endorses candidates in state legislative races.

INDIANA

Hoosier Environmental Council, 1002 E. Washington, Suite 300, Indianapolis, IN 46202. Phone: (317) 685-8800.
Can provide details on state environmental issues.

Sierra Club, 6140 N. College Ave., Indianapolis, IN 46220. Phone: (317) 253-2687.
Produces a scorecard rating legislators on 6-10 key environmental bills; endorses candidates in legislative races.

IOWA

Iowa Citizen Action Network, 415 10th Street, Des Moines, IA 50309. Phone: (515) 244-9311.
Plans to publish a legislative scorecard in time for the fall 1992 election.

KANSAS

Kansas Sierra Club, 2935 S. Seneca, Wichita, KS 67217. Phone: (316) 683-8492.
Produces an environmental legislative scorecard and endorses candidates in state legislative races.

KENTUCKY

Kentucky Sierra Club, c/o Dave Stawicki, 2004 Ritt Court, Lexington, KY 40504.
Makes endorsements in state legislative races.

LOUISIANA

Louisiana Environmental Action Network, P.O. Box 66323, Baton Rouge, LA 70896. Phone: (504) 928-1315.
Produces an environmental "report card" on local legislators.

MAINE

Maine League of Conservation Voters, P.O. Box 5271, Augusta, ME 04332. Phone: (207) 871-7793.
Legislative scorecard published biannually.

Maine People's Alliance, P.O. Box 17534, Portland, ME 04101. (207) 761-4400. *Publishes a scorecard; endorses legislative candidates.*

MARYLAND

League of Conservation Voters, 1028 Old Bay Ridge Road, Annapolis, MD 21403. Phone: (410) 267-0716.
Publishes a voting chart on members of the state assembly.

MASSACHUSETTS

Environmental Lobby of Massachusetts, 3 Joy Street, Boston, MA 02108. Phone: (617) 742-2553.
Publishes a legislative newsletter.

MICHIGAN

Michigan Environmental Council, 115 W. Allegan, Suite 10B, Lansing, MI 48933. Phone: (517) 487-9539.
Publishes a yearly Action Agenda and Update *on environmental issues in the legislature.*

Michigan Sierra Club, 115 W. Allegan, Suite 10B, Lansing, MI 48933. Phone: (517) 484-2372.
Publishes a voter's guide; endorses candidates in state legislative races.

MINNESOTA

Minnesota League of Conservation Voters, P.O. Box 580095, Minneapolis, MN 55458-0095.
Publishes a legislative scorecard based on a consensus of state environmental lobbyists.

Sierra Club, North Star Chapter, 1313 5th Street SE, Suite 323, Minneapolis, MN 55414. (612) 379-3853.
Publishes a legislative voting card and endorses candidates in state races.

MISSISSIPPI

Environmental Coalition of Mississippi, P.O. Box 31292, Jackson, MS 39286.
Publishes a legislative scorecard.

Sierra Club, 921 North Congress, Jackson, MS 39202. Phone: (601) 352-1026.
Endorses candidates in state races.

MISSOURI

Missouri Audubon Council, c/o Charlie Callison, 403 Castle Drive, Jefferson City, MO 65109.
After each session, publishes a summary of state environmental bills.

Sierra Club, Ozark Chapter, P.O. Box 364, Jefferson City, MO 65102. Phone: (314) 645-1019.
Has published a state legislative scorecard for the past two years.

MONTANA

Montana Environmental Information Center, P.O. Box 1184, Helena, MT 59624. Phone: (406) 443-2520.
Publishes a biweekly update called "The Capitol Monitor" during the legislative session. Also available: An environmental voting record.

Montana Sierra Club, 78 Konley Dr., Kalispell, MT 59901.

NEBRASKA

Nebraska Audubon Council, c/o Dave Sands, 7700 SW 27th St., Lincoln, NE 68523.
Publishes information on state environmental legislation each session.

Nebraska Coalition for the Environment, 1640 L Street, Suite G, Lincoln, NE 68508. Phone: (402) 476-0444.
Publishes a "Green Sheet" with a legislative environmental agenda for each session.

NEVADA

Nevada Sierra Club, P.O. Box 8096, Reno, NV 89507. Phone: (702) 323-3162.
Endorses candidates in state races.

NEW HAMPSHIRE

New Hampshire League of Conservation Voters, 3 Market Square, Portsmouth, NH 03801. Phone: (603) 627-8935.
Works with several statewide groups to produce a legislative scorecard.

New Hampshire Citizen Action, 10 Ferry Street, Box 319, Concord, NH 03301. Phone: (603) 225-2097.
Publishes a legislative scorecard on environmental and economic issues; endorses candidates for the state legislature and governor's office.

NEW JERSEY

New Jersey Environmental Federation, 46 Bayard Street, New Brunswick, NJ 08901. Phone: (908) 846-4224.
Publishes a voter scorecard and endorses candidates in state legislative races.

New Jersey Environmental Lobby, 204 West State Street, Trenton, NJ 08608. Phone: (609) 396-3774.
Publishes a quarterly newsletter describing major environmental bills and how members of the legislature voted.

NEW MEXICO

New Mexico Conservation Voters Alliance, P.O. Box 40497, Albuquerque, NM 87196.
Publishes a legislative report and scorecard called "Green Sweep Report."

New Mexico Sierra Club, 207 San Pedro Ave. NE, Albuquerque, NM 03301. Phone: (505) 265-5506.
Endorses candidates in state races.

NEW YORK

New York Environmental Planning Lobby, 353 Hamilton Street, Albany, NY 12210. Phone: (518) 462-5526.
Publishes an annual Environmental Voter's Guide.

NORTH CAROLINA

N.C. League of Conservation Voters, P.O. Box 12462, Raleigh, NC 27605.
Publishes a local voter chart and endorsement sheet.

NORTH DAKOTA

Clean Water Action, 118 N. Broadway, 211 Black Building, Fargo, ND 58102. Phone: (701) 235-5431.
Publishes an endorsement sheet on local legislators.

OHIO

Ohio Environmental Council, 145 N. High Street, Suite 509, Columbus, OH 43215. Phone: (614) 224-4900.
Endorses candidates in state races.

OKLAHOMA

Sierra Club, P.O. Box 60882, Oklahoma City, OK 73416. Phone: (405) 721-5486.
Publishes a scorecard for the state legislature.

OREGON

Oregon League of Conservation Voters, 520 Southwest 6th, Suite 701, Portland, OR 97204. Phone: (503) 224-4011.
Publishes a biannual legislative scorecard.

Sierra Club, 1413 Southeast Hawthorne Blvd., Portland, OR 97214. Phone: (503) 238-0442.
Endorses candidates in state races.

PENNSYLVANIA

Pennsylvania Sierra Club, 600 N. 2nd St., P.O. Box 663, Harrisburg, PA 17108. (717) 232-0101.
General information and advice.

RHODE ISLAND

Ocean State Action, 99 Bald Hill Road, Cranston, RI 02920. Phone: (401) 463-5368.
Publishes a legislative scorecard on the environment and other issues.

SOUTH CAROLINA

South Carolina Sierra Club, 1314 Lincoln St., Columbia, SC 29211. Phone: (803) 256-8487.
General information and advice.

SOUTH DAKOTA

South Dakota Resources Council, c/o David Nelson, Box 7020, University Station, South Dakota State University, Brookings, SD 57007.
Publishes an analysis of each legislative session in its newsletter.

TENNESSEE

Tennessee Sierra Club, P.O. Box 158472, Nashville, TN 37215.
Publishes a legislative newsletter and scorecard; endorses candidates in state races.

TEXAS

Texas Sierra Club, P.O. Box 1931, Austin, TX 78767. Phone: (512) 477-1729.
Publishes a legislative voting guide and newsletter with endorsements.

Public Citizen of Texas, 1205 Nueces Street, Austin, TX 78701. Phone: (512) 477-1155.
General information and advice.

UTAH

Utah Sierra Club, 177 East 900 South, Suite 102, Salt Lake City, UT 84111. Phone: (801) 363-9621.
Publishes a legislative scorecard.

Utah Outdoor Alliance, 50 South Main #710, Salt Lake City, UT 84144.
Publishes a legislative scorecard and voter's guide.

VERMONT

Vermont PIRG, 43 State Street, Montpelier, VT 05602. Phone: (802) 223-5221.
Publishes a legislative voting record and scorecard.

VIRGINIA

Virginia Environmental Network, 1001 East Broad Street, Richmond, VA 23219. Phone: (804) 644-0283.
Publishes an environmental legislative guide.

WASHINGTON

Olympia Sierra Club, 2631 12th Court, SW, Suite A, Olympia, WA 98502. Phone: (206) 754-2386.
Endorses candidates in state legislative races.

Washington Environmental Political Action Committee, P.O. Box 85914, Seattle, WA 98145-1194. Phone: (206) 783-4045.
Publishes a legislative scorecard and endorses candidates in state races.

WEST VIRGINIA

West Virginia Citizen Action Group, 1324 Virginia Street East, Charleston, WV 25301. Phone: (304) 346-5891.
Publishes a legislative scorecard and endorses candidates in state races.

Sierra Club, P.O. Box 4142, Morgantown, WV 26504.
Publishes a legislative scorecard; endorses candidates.

WISCONSIN

Wisconsin Environmental Decade, Green Vote, 122 State Street, Suite 200, Madison, WI 53703. Phone: (608) 251-7020.
Publishes a legislative scorecard and makes endorsements.

John Muir Sierra Club, 111 King Street, Madison, WI 53703.
Phone: (608) 256-0565.
Makes endorsements in state races.

WYOMING

Sierra Club Political Action Committee, 615-1/2 Clark Street,
Laramie, WY 82070.
Endorses candidates for state legislative offices.

Wyoming Outdoor Council, 201 Main Street, Lander, WY
82520. Phone: (307) 332-7031.
*Publishes a summary of environmental bills and a legislative voting
record.*

ELECTION RESOURCES

Here are some national environmental organizations committed to helping voters understand the issues. Some work directly with pro-Earth candidates. Write or call them for a complete description of what they have and how they can help you identify pro-Earth candidates.

Environmental Action, 6930 Carroll Ave., #600, Takoma Park, MN 20912. (301) 891-1100. *What's available: legislative summaries on solid waste issues. You can also get their publication "A Brighter Future," which tracks state legislation and regulations, for $10.*

Environmental Defense Fund, 1875 Connecticut Ave., 10th Floor, Washington, DC 20009. (202) 387-3500. *Doesn't endorse candidates but has general information on issues.*

Friends of the Earth, 218 D St., SE, Washington, DC 20003. (202) 544-2600. *Has a fact sheet called "Pressuring Politicians" and legislative "Action Alerts" for members. Membership is $25/year.*

Greenpeace, 1436 U St., NW, Washington, DC 20009. (202) 462-1177. *Voters can join their Activist Network to receive information on legislation and letter-writing campaigns (it's free).*

League of Conservation Voters, 1707 L St., NW, Suite 550, Washington, DC 20036. (202) 785-8683. *Their "National Environmental Scorecard"—a longer version of the "Scorecard" section of this book—is available for $6. "Presidential Profiles" summarizes the presidential candidates' records on the environment; it's $2. Candidate information for all federal elections is available.*

National Audubon Society, 666 Pennsylvania Ave., SE, Washington, DC 20003. (202) 547-9009. *What's available: Materials for activists including "Activist Toolkits," information packets on wetlands, the Arctic Refuge, and ancient forests ($6 each).*

Sierra Club/Political Action Committee, 408 C St., NE, Washington, DC 20002. (202) 547-1141. *Endorses candidates and provides general information for voters. Has a publication called* Election Action Bulletin *for members. Membership is $35/year.*